New Ideas Concerning Black Holes and the Universe

Edited by Eugene Tatum

Published in London, United Kingdom

IntechOpen

Supporting open minds since 2005

New Ideas Concerning Black Holes and the Universe
http://dx.doi.org/10.5772/intechopen.75238
Edited by Eugene Tatum

Contributors
Miroslav Pardy, Golden Gadzirayi Nyambuya, Mikael Souto Maior De Sousa, Anderson Alves De Lima, Aloke Kumar Sinha, Eugene Tatum

First published in London, United Kingdom, 2020 by IntechOpen
IntechOpen is the global imprint of INTECHOPEN LIMITED, registered in England and Wales, registration number: 11086078, 7th floor, 10 Lower Thames Street, London, EC3R 6AF, United Kingdom
Printed in Croatia

British Library Cataloguing-in-Publication Data
A catalogue record for this book is available from the British Library

Additional hard and PDF copies can be obtained from orders@intechopen.com

New Ideas Concerning Black Holes and the Universe
Edited by Eugene Tatum
p. cm.
Print ISBN 978-1-83968-541-5
Online ISBN 978-1-83968-542-2
eBook (PDF) ISBN 978-1-83968-543-9

For EU product safety concerns:
IN TECH d.o.o., Prolaz Marije Krucifikse Kozulić 3, 51000 Rijeka, Croatia, info@intechopen.com or visit our website at intechopen.com.

Meet the editor

Dr. Tatum is today's foremost expert on the apparent links between black hole physics and cosmology. This has a longstanding tradition dating back to the Hawking–Penrose conjecture that time-reversal of universal expansion from a 'singularity' must have physical features remarkably similar to black hole collapse to a 'singularity'. Following the initial Planck Collaboration reports, Dr. Tatum published two seminal papers comparing their observational findings to the Hawking–Penrose conjecture. Since then, he has lead-authored more than 20 peer-reviewed scientific publications on Flat Space Cosmology (FSC). This is believed to be the first predictive cosmology model using Hawking's singularity theorem as its founding principle. FSC is now the primary competitor to the inflationary 'concordance model' of cosmology.

Contents

Preface

The ideas presented in this book are new scientific theories based largely upon a spate of very recent astronomical observations. These theories include: a cosmological model that appears to be superior, in many respects, to the inflationary 'concordance model'; proposed thermal stability criteria for a generic quantum black hole; theoretical constraints concerning black hole binary graviton emissions; theoretical effects of abelian vortices on space-time; and a proposed solution to the mystery of the observed asymmetry between universal matter and antimatter.

Beginning with NASA's 1990 deployment of the Hubble Space Telescope, an ingenious variety of observational platforms (Earth-based and space-based) have been rapidly approaching the theoretical limits of spectral observation. Thus, there is a widespread consensus that we are now in a 'golden age' of astronomy and cosmology. Unless radically new laws of physics are still left to be discovered, it is doubtful that scientists of the future will be able to look much farther into space and time than our current generation.

Nevertheless, even if that proves to be the case, we will long continue to refine our thinking and understanding of what we are now observing. Vigorous scientific debate about the inner workings of black holes and the precise conditions of the early universe will likely continue unabated. This is a good thing. The new ideas presented in this book have been selected in order to inspire others that, regardless of the impending limits of observation, the scientific creative process will continue.

Eugene Tatum
Independent Researcher,
Bowling Green, Kentucky, USA

Chapter 1

Introductory Chapter: Black Holes, The Singularity Problem, and The Universe

Eugene Terry Tatum

1. Black holes

Black holes are arguably the most fascinating and mysterious objects in the known universe. Our fascination is heightened by the recent imaging, for the first time, of the event horizon of a galactic supermassive black hole. This was a monumental feat of ingenuity and engineering on the part of contemporary astronomers, mathematicians, computer programmers, and physicists. And it was, equally, a monumental achievement of the human mind, dating back to the work of Albert Einstein in the early days of the twentieth century.

As first recognized by German physicist Karl Schwarzschild in 1916, Einstein's general theory of relativity predicts the existence of a particularly strange phenomenon occurring inside the radius of any celestial object whose mass becomes equal to $rc^2/2G$. It was soon apparent that an object satisfying the Schwarzschild metric should not allow anything, including light, to escape from within this "horizon" radius.

2. Singularity problem

Stranger still, it eventually became apparent that such an object had no obvious means of stabilizing itself at any given radius of gravitational collapse, *no matter how small!* Effectively, an empty black "hole" would be created inside the Schwarzschild radius, with the exception of an infinitely small and infinitely dense "singularity" at the hole's geometric center. This was a strange prediction of general relativity that even Einstein could not accept. After all, what could "infinitely small" and "infinitely dense" even mean?!

Infinity may be an acceptable concept to a mathematician, but physicists tend to abhor the idea of real objects with infinite properties. Accordingly, we would prefer to invoke real or imagined principles of quantum physics in order to avoid the conundrum presented by an infinite singularity. This has become the primary inspiration for developing theories of quantum gravity collectively known as loop quantum gravity. It is also a primary motivation for the intrinsically beautiful, although somewhat abstract, hyperdimensional and mathematically complex collection of string theories. Unfortunately, there is not, as of yet, a fully coherent and provable theory of quantum gravity upon which we can hope to understand the inner workings of a black hole, to say nothing of the gravitational conditions at the inception of the universe. This singularity problem is just one of the many puzzling things about black holes and the very early universe, some of which are addressed in this book.

IntechOpen

3. Universe

In the 1960s, English mathematical physicists Roger Penrose and Stephen Hawking cleverly extended the black hole singularity problem to our expanding universe. They proved that a backward time extrapolation of the expansion should inevitably lead to the same problem implied at the geometric center of a Schwarzschild black hole. Thus, allowing for the time symmetry of general relativity, the implication of their work [1–3] was that our universe has this feature, *and perhaps others*, in common with a nonrotating, electrically neutral, black hole. Whether the universe could be a time-reversed black hole-like object (sometimes referred to as a "white hole") has been a subject of vigorous debate over the last 50 years [4–11].

When one rearranges the Schwarzschild formula, it is readily apparent that the M/r ratio (the ratio of the gravitational mass to Schwarzschild's horizon radius) of an equilibrated, nonrotating, electrically neutral black hole must equal $c^2/2G$. This ratio (approximately 6.73×10^{26} kg/m in metric units) is effectively a constant of nature incorporating two of the most fundamental constants of nature, Maxwell's speed of light and Newton's gravitational constant. Furthermore, one could make a strong argument that this mathematical relationship is a reliable *signature* of a black hole or black hole-like object.

One of the features surprisingly in common between Schwarzschild black holes and the observable universe has been documented fairly recently and has been a subject of great interest to myself and others. A series of astronomical observations since the early 1990s [12–15] allow one to calculate a reasonably accurate M/r ratio value for the observable universe. If so inclined, the interested reader can skip the foundational references indicated and simply look up the relevant numbers on the Wikipedia link entitled "observable universe." The mass M of the observable universe is now estimated to be 1.5×10^{53} kg, and the radius r of the observable universe is now estimated to be 4.4×10^{26} m. One can readily see that this implies a current M/r ratio value for the observable universe of approximately 3.4×10^{26} kg/m. Consequently, *the M/r ratio values for an equilibrated Schwarzschild black hole and the expanding observable universe are of the same order of magnitude*. One should let that sink in.

4. A useful model

The above calculation made from recently published observations was unavailable to an earlier generation of physicists. This new result, in combination with the time-symmetric properties of general relativity enshrined within Hawking's singularity theorem, provides an excellent starting point for exploring the heuristic cosmology model of the universe I present in this book. It is believed to be the first cosmology model of its kind, namely, that which solely incorporates in its assumptions reasonable speculations about black hole time reversal. Thus, Hawking's singularity theorem is the founding principle of this model which I call flat space cosmology.

So far, this model appears to be quite accurate with respect to correlations between its embedded predictions at any particular point in cosmic time and a variety of astronomical observations. It is my hope that the reader will begin with this chapter and be inspired to study the model further. The competing "concordance model" incorporating inflationary cosmology may not be the last word after all!

5. A sense of wonder

As the reader delves further into this book, it is also my hope that he or she will have a sense of wonder for how far we have come in understanding black holes and the rules governing the expansion of our universe. That so much of our universe is actually comprehensible was a wonder to even Albert Einstein. However, as the new ideas presented in this book clearly show, there are still many creative avenues for further exploration.

Author details

Eugene Terry Tatum
Independent Researcher, Bowling Green, Kentucky, USA

*Address all correspondence to: ett@twc.com

References

[1] Penrose R. Gravitational collapse and space-time singularities. Physical Review Letters. 1965;**14**(3):57-59

[2] Hawking S. Properties of expanding universes [PhD.5437]. Cambridge University Library; 1966. Available from: https://cudl.lib.cam.ac.uk/view/MS-PHD-05437/134

[3] Hawking S, Penrose R. The singularities of gravitational collapse and cosmology. Proceedings of the Royal Society of London A. 1970;**314**:529-548

[4] Pathria RK. The universe as a black hole. Nature. 1972;**240**(5379):298299

[5] Poplawski N. Radial motion into an Einstein-Rosen bridge. Physics Letters B. 2010;**687**(2-3):110-113. DOI: 10.1016/j.physletb.2010.03.029

[6] Seshavatharam UVS. Physics of rotating and expanding black hole universe. Progress in Physics. 2010;**2**:7-14

[7] Cowen R. Quantum bounce could make black holes explode. Interview with Carlo Rovelli. 2014. Available from: www.nature.com/1.15573

[8] Tatum ET. Could our universe have features of a giant black hole? Journal of Cosmology. 2015;**25**:13061-13080

[9] Tatum ET. How a black hole universe theory might resolve some cosmological conundrums. Journal of Cosmology. 2015;**25**:13081-13111

[10] Haggard HM, Rovelli C. Quantum-gravity effects outside the horizon spark black to white hole tunneling. Physical Review D. 2015;**92**:104020

[11] Laguipo A. Black holes may be possible portals to another universe: Stephen Hawking. Interview article. 2016. Available from: www.techtimes.com/articles/153308

[12] Bars I, Terning J. Extra Dimensions In Space And Time. Springer; 2009. p. 27. ISBN 978-0-387-77637-8

[13] Planck Collaboration. Planck 2015 results. XIII. Cosmological parameters. Astronomy & Astrophysics. 2016;**594**:A13. arXiv:1502.01589. DOI: 10.1051/0004-6361/201525830

[14] Lineweaver C, Davis TM. Misconceptions about the big bang. Scientific American. 2005;**292**(3):36-45. DOI: 10.1038/scientificamerican0305-36

[15] Bennett CL et al. Seven-year Wilkinson microwave anisotropy probe (WMAP) observations: Are there cosmic microwave background anomalies? Astrophysical Journal Supplement. 2011;**192**(2):17. arXiv:1001.4758. DOI: 10.1088/0067-0049/192/2/17

Chapter 2

A Heuristic Model of the Evolving Universe Inspired by Hawking and Penrose

Eugene Terry Tatum

Abstract

A heuristic model of universal expansion is presented which uses, as its founding principle, Stephen Hawking's singularity theorem. All assumptions of this model are intrinsically linked to Hawking's theorem and its implications with respect to the time-symmetric properties of general relativity. This is believed to be the first mathematical model constructed in such a way, and it is remarkably accurate with respect to current astrophysical observations. This model's apparent superiority to standard inflationary cosmology is emphasized throughout, including its accurate derivations of the observed Hubble parameter value and CMB anisotropy. The model definition of cosmic entropy not only correlates the observed temperature anisotropy but also may have implications for resolving the cosmological constant problem and the mystery of dark energy. Moreover, the model has a temperature curve which is more favorable for the remarkably early formation of quasars and galaxies. Possible deep connections to Verlinde's "emergent gravity" theory are also discussed.

Keywords: flat space cosmology, cosmology theory, cosmic inflation, dark energy, cosmic flatness, CMB anisotropy, cosmic entropy, black holes, cosmic dawn, $R_h = ct$ model

1. Introduction and background

A heuristic mathematical model of the evolving universe, for the purpose of this chapter, is one which tracks its *global* parameters (Hubble parameter, radius, mass, energy, entropy, average temperature, temperature anisotropy, etc.) as a function of cosmic time. For it to be useful, such a model should be consistent with everything we currently observe about the universe as a global object and extend these parameters indefinitely into the past and future. In assembling such a model, it is particularly useful to start with a founding principle on which some or, preferably, all of the starting assumptions can be based. For this particular model, the founding principle is based upon the groundbreaking work of Roger Penrose [1] and Stephen Hawking [2, 3] concerning the similar theoretical nature of astrophysical and cosmological singularities. This founding principle is Hawking's singularity theorem.

Hawking's singularity theorem implies that our universe, following time-symmetric properties of general relativity, could be treated mathematically as if it were a cosmological black hole-like object moving *backward in time* (i.e., *expanding from* a singularity state as opposed to *collapsing to* a singularity state). Unfortunately, although Hawking's theorem was rigorously logical, he never actually put together a predictive mathematical cosmological model based upon his theorem. *What is presented in this chapter is believed to be the first such model.*

This author was sufficiently intrigued by the potential implications of Hawking's singularity theorem that he teamed up with two Indian physicists (U.V.S. Seshavatharam and S. Lakshminarayana) in 2015 to publish the seminal papers [4–6] on this model. For reasons to be discussed below, this model is called "flat space cosmology" (FSC). The current five basic assumptions of FSC are presented below.

2. The five basic assumptions of flat space cosmology

1. The cosmic model is an ever-expanding sphere such that the cosmic horizon always translates at speed of light c with respect to its geometric center at all times t. The observer is operationally defined to be at this geometric center at all times t.

2. The cosmic radius R_t and total mass M_t follow the Schwarzschild formula $R_t \cong 2GM_t/c^2$ at all times t.

3. The cosmic Hubble parameter is defined by $H_t \cong c/R_t$ at all times t.

4. Incorporating our cosmological scaling adaptation of Hawking's black hole temperature formula, at any radius R_t, cosmic temperature T_t is inversely proportional to the geometric mean of cosmic total mass M_t and the Planck mass M_{pl}. R_{pl} is defined as twice the Planck length (i.e., as the Schwarzschild radius of the Planck mass black hole). With subscript t for any time stage of cosmic evolution and subscript pl for the Planck scale epoch and incorporating the Schwarzschild relationship between M_t and R_t,

$$k_B T_t \cong \frac{\hbar c^3}{8\pi G\sqrt{M_t M_{pl}}} \cong \frac{\hbar c}{4\pi\sqrt{R_t R_{pl}}}$$

$$\left\{\begin{array}{ll} M_t \cong \left(\frac{\hbar c^3}{8\pi G k_B T_t}\right)^2 \dfrac{1}{M_{pl}} & \text{(A)} \\ R_t \cong \dfrac{1}{R_{pl}}\left(\dfrac{\hbar c}{4\pi k_B}\right)^2\left(\dfrac{1}{T_t}\right)^2 & \text{(B)} \\ R_t T_t^2 \cong \dfrac{1}{R_{pl}}\left(\dfrac{\hbar c}{4\pi k_B}\right)^2 & \text{(C)} \\ t \cong \dfrac{R_t}{c} & \text{(D)} \end{array}\right\} \quad (1)$$

5. Total cosmic entropy follows the Bekenstein-Hawking black hole entropy formula [7, 8]:

$$S_t \cong \frac{\pi R_t^2}{L_p^2} \quad (2)$$

The rationale for these basic assumptions is closely tied to Hawking's singularity theorem as it might pertain to a time-reversed Schwarzschild cosmological black hole-like object. From the centrally located observer's point of view, outwardly moving photons traveling along geodesics at the cosmic boundary (i.e., the fastest-moving "particles" of the expansion) are infinitely redshifted and thus define the observational event horizon. Therefore, as given in assumption 3, the truly *global* Hubble parameter value can always be defined as speed of light c divided by the ever-increasing Schwarzschild radius R_t. While the first equation of assumption 4 closely resembles Hawking's black hole temperature formula, it is modified so that cosmological mass scales in Planck mass units. This is thought to be more appropriate for a scaling cosmological model, as opposed to the relatively static thermodynamics of an astrophysical (i.e., stellar) black hole.

As described in some detail in the seminal FSC papers, the first three assumptions allow for perpetual Friedmann's critical density (i.e., perpetual global spatial flatness) of the expanding FSC cosmological model from its inception. It should be emphasized that these assumptions were not adopted for this particular purpose. However, this unexpected and fortuitous outcome is perhaps the most important feature of this model. By dividing the Schwarzschild mass (defined in terms of cosmic radius R_o) by the spherical volume and substituting c^2/R_o^2 with H_o^2, Friedmann's critical mass density $\rho_0 = \frac{3H_0^2}{8\pi G}$ is achieved for any given moment of theoretical observation (hence the subscript "o") in cosmic time. So, *perpetual Friedmann's critical density and global spatial flatness from inception is a fundamental feature of the FSC model. Our model was named for this important feature.*

This perpetual spatial flatness feature, as well as the finite properties of light-speed expansion of the cosmic horizon, obviates the need for an inflationary solution to the cosmological "flatness problem" and the "horizon problem." It also avoids the disturbing and incredible "infinite multiverse" implications inherent within inflationary cosmic models. The problems of the required new physics of the "inflaton" field, and of the "past-incomplete" nature [9] of inflationary models, are also avoided in the FSC model. Many of these differentiating features of FSC with respect to standard inflationary models were discussed at length in a recent FSC summary paper [10].

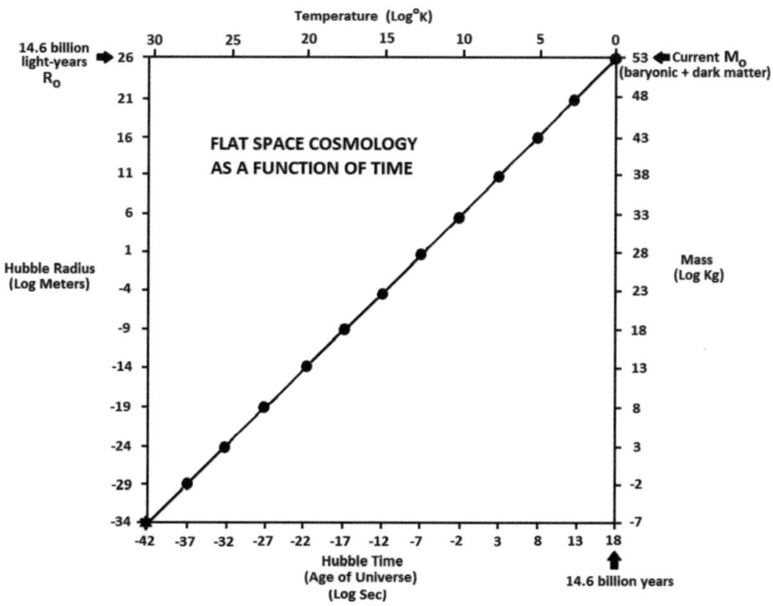

Figure 1.
Cosmic radius, temperature, and mass as a function of cosmic time.

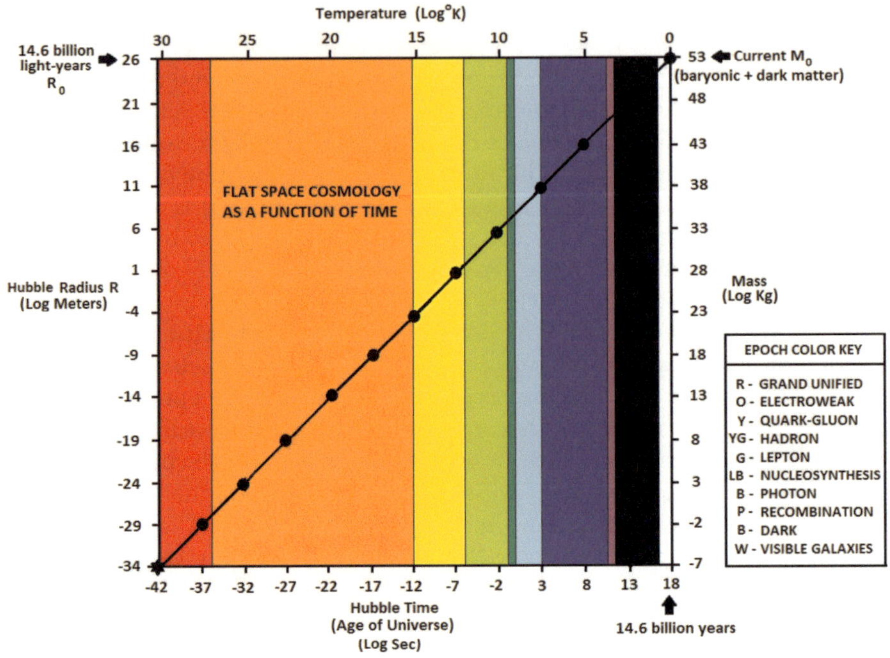

Figure 2.
Particle physics epochs as a function of cosmic time and temperature.

Based upon the relations proceeding from the top equation of assumption 4, and the model Hubble parameter definition of assumption 3, an FSC log graph can be presented in **Figure 1**.

A color-coded overlay of cosmic epochs evolving from the Planck scale epoch, as believed to be the case from particle physics experiments and quantum field theory, is presented in **Figure 2**.

In both figures, there is a tight correlation between cosmic temperature and time elapsed since the Planck scale epoch (not shown) at approximately the 10^{-43} s mark of cosmic expansion.

3. FSC correlations with astronomical observations

The following temperature-dependent cosmological parameters can be easily calculated in the FSC model. The only free parameter in any of these equations is the cosmic temperature. Furthermore, by incorporating the values of T_0, $\hbar$, c, G, k_B, L_p, and π to as many decimal places as known, any of these FSC parameters can be shown to closely match astronomical observations:

$$R \cong \frac{\hbar^{3/2}c^{7/2}}{32\pi^2 k_B^2 T^2 G^{1/2}} \qquad R_0 \cong \frac{\hbar^{3/2}c^{7/2}}{32\pi^2 k_B^2 T_0^2 G^{1/2}} \tag{3}$$

$$H \cong \frac{32\pi^2 k_B^2 T^2 G^{1/2}}{\hbar^{3/2}c^{5/2}} \qquad H_0 \cong \frac{32\pi^2 k_B^2 T_0^2 G^{1/2}}{\hbar^{3/2}c^{5/2}} \tag{4}$$

$$t \cong \frac{\hbar^{3/2}c^{5/2}}{32\pi^2 k_B^2 T^2 G^{1/2}} \qquad t_0 \cong \frac{\hbar^{3/2}c^{5/2}}{32\pi^2 k_B^2 T_0^2 G^{1/2}} \tag{5}$$

$$M \cong \frac{\hbar^{3/2} c^{11/2}}{64\pi^2 k_B^2 T^2 G^{3/2}} \qquad M_0 \cong \frac{\hbar^{3/2} c^{11/2}}{64\pi^2 k_B^2 T_0^2 G^{3/2}} \tag{6}$$

$$Mc^2 \cong \frac{\hbar^{3/2} c^{15/2}}{64\pi^2 k_B^2 T^2 G^{3/2}} \qquad M_0 c^2 \cong \frac{\hbar^{3/2} c^{15/2}}{64\pi^2 k_B^2 T_0^2 G^{3/2}} \tag{7}$$

Current parameters are calculated in the right-hand column. The currently observed cosmic temperature value T_0 = 2.72548 K. Accordingly, the theoretical current FSC Hubble parameter value at this temperature is.

$$H_0 = 2.167862848658891 \text{ x } 10^{-18}\ \text{s}^{-1}\ \left(66.89325791854758\ \text{km.s}^{-1}.\text{Mpc}^{-1}\right)$$

This derived theoretical global H_o value fits the 2018 Planck Collaboration observational global H_o value of 67.36 +/− 0.54 km.s^{-1}.Mpc^{-1} (68% confidence interval for TT, TE, EE + lowE + lensing) [11] and the DES 2018 H_o value of 67.77 $\pm$ 1.30 km s^{-1}.Mpc^{-1} (SN + BAO) [12]. Since the Planck observational value was obtained *partially* with the aid of extraordinarily precise observations of the CMB black body radiation spectrum, this may be as close as we can come in the foreseeable future to a truly *global* Hubble parameter measurement. And yet, the above theoretical H_o calculation is based *solely* upon this one carefully measured free parameter: T_0 = 2.72548 K. This is a remarkable result!

Therefore, one should have great confidence that the following cosmological parameters incorporating the FSC-derived H_o value are also highly accurate:

$$t_0 \cong \frac{1}{H_0} = 4.61283794 \text{ x } 10^{17}\ \text{s}\ \left(14.61694684 \times 10^9\ \text{sidereal years}\right)$$

(multiplying by 1 sidereal year per 3.155814954 $\times$ 10^7 s)

This value is simply the reciprocal of the above-derived Hubble parameter value, as one would expect for the perpetually spatially flat FSC cosmic model in comparison with the standard inflationary model. For reasons not elaborated here, *any* inflationary model would be *expected* to calculate a slightly younger cosmic age. 13.8 billion years is now consensus for the standard inflationary model:

$$R_0 \cong \frac{c}{H_0} = 1.38289402 \text{ x } 10^{26}\ \text{m}\ \left(14.617201\ \times\ 10^9\ \text{light} - \text{years}\right)$$

(multiplying by 1 Julian light-year per 9.4607304725808 $\times$ 10^{15} m)

This current cosmic radius value correlates with current cosmic time by $R_o = ct_o$. Therefore, FSC is a $R_h = ct$ cosmological model. Later discussion in this chapter will focus on the extremely good statistical fit between $R_h = ct$ models and the accumulated Type Ia supernovae light curve data purported to "prove" the existence of cosmic acceleration:

$$Vol_0 = \frac{4\pi}{3}\left(\frac{c}{H_0}\right)^3 = 1.10778456 \times 10^{79}\ \text{m}^3$$

$$M_0 = \frac{c^3}{2GH_0} = 9.31126529 \times 10^{52}\ \text{kg}$$

This total mass number can be compared very favorably to a rough estimate made from astronomical observations. The visible matter consists of roughly 100 billion galaxies averaging roughly 100 billion stars each, of average star mass equal to roughly 1.4 $\times$ 10^{30} kg (70% of solar mass), totaling to roughly 1.4 $\times$ 10^{52} kg.

The 2015 Planck Collaboration report indicates a universal matter ratio of approximately 5.5 parts dark matter to 1 part visible (baryonic) matter. This brings the total estimated matter in the observable universe to approximately 9.1×10^{52} kg. A recent study [13] of average mass density of intergalactic dust gives a value of approximately 10^{-30} kg.m^{-3}. Since this is approximately 1 part intergalactic dust to 1000 parts galactic and perigalactic matter, intergalactic dust does not appreciably modify the estimated total observational mass of matter given above. Accordingly, this observational estimate is remarkably close to the above FSC theoretical calculation of total cosmic mass attributed to positive energy (i.e., gravitationally attractive) matter.

According to the FSC Friedmann equations (referenced below), the positive matter mass-energy is equal in absolute magnitude, and opposite in sign, to the negative (dark) energy at all times. This is a 50/50 percentage ratio as opposed to the approximately 30/70 ratio implied by yet unproven, and supposedly dark energy-dominating, cosmic *acceleration*. However, without definitively proving cosmic acceleration, standard inflationary cosmology cannot claim this 30/70 ratio! (Please see the discussion and relevant references in the last two paragraphs of this section):

$$M_0c^2 = \frac{c^5}{2GH_0} = 8.3685479 \text{ x } 10^{69} \text{ J}$$

$$\rho_0 = \frac{3H_0^2}{8\pi G} = 8.40530333 \text{ x } 10^{-27} \text{ kg.m}^{-3} \text{ (critical mass density)}$$

This closely approximates the observational cosmic mass density calculation of critical density:

$$\rho_0 c^2 = \frac{3H_0^2c^2}{8\pi G} = 7.554309896 \text{ x } 10^{-10} \text{ J.m}^{-3} \text{ (critical mass – energy density)}$$

This closely approximates the observational cosmic mass-energy density and the observational vacuum energy density. They are equal in absolute magnitude, and opposite in sign, in FSC.

A recent paper [14] has integrated the FSC model into the Friedmann equations containing a Lambda Λ cosmological term. Thus, FSC has been shown to be a scalar dynamic Λ dark energy model of the wCDM type (wherein equation of state term w is always equal to -1.0). Furthermore, it is well-known that a sufficiently realistic $R_h = ct$ model, such as FSC, can fit within the tightest constraints of the Supernova Cosmology Project (SCP) data. The following open-source graph (**Figure 3**) from the SCP is offered as proof [15].

One can readily see (by the "flat" line intersection) that a realistic spatially flat universe model such as FSC is an excellent fit with all such SCP observations to date.

Currently, there is no certainty about the percentage of the critical density which is attributable to dark matter. Those with knowledge of the observational studies of the ratio of dark matter to visible matter realize the difficulty of determining a precise *co-moving* value for this ratio at the present time. Galactic and perigalactic distributions of dark matter can be surprisingly variable, as evidenced by the 29 March 2018 report in *Nature* [16] of a galaxy apparently completely lacking in dark matter! Although the 2015 Planck Collaboration consensus is a large-scale approximate ratio of 5.5 parts dark matter to 1 part visible matter, this can only be considered as a rough estimate of the actual *co-moving* ratio, particularly if this ratio varies significantly over cosmic time. A 9.2-to-1 actual ratio in approximately co-moving galaxies (i.e., those within about 100 million light-years of the Milky

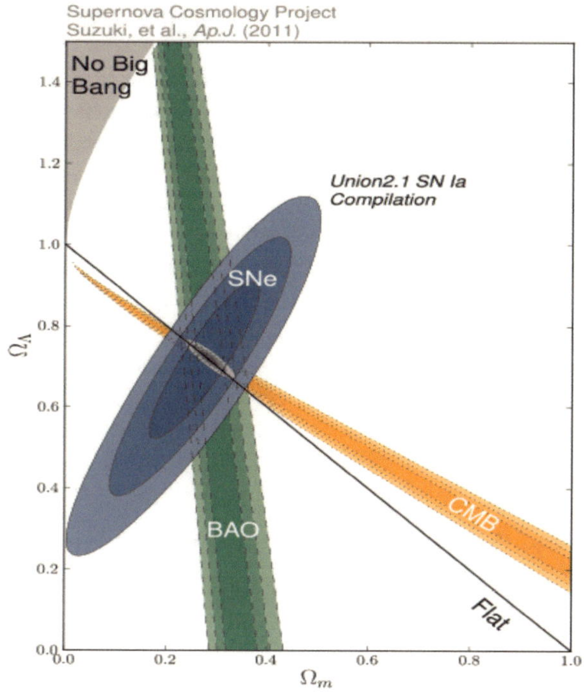

Figure 3.
SCP supernovae, BAO, and CMB data.

Way galaxy) remains a possibility and would change the ratio of total matter mass-energy to dark energy to essentially unity (i.e., 50% matter mass-energy and 50% dark energy). Thus, the intersection zone of tightest constraints shown in **Figure 3** should then correlate with 0.5 Omega_ m and 0.5 Omega_Λ. This is one of several important testable predictions discriminating the FSC model from the standard inflationary cosmology model. Precise measurements of approximately co-moving galaxies are in order, for comparison with the CMB observational Planck Collaboration result.

The question of dark energy density dominance over total matter energy density remains in doubt, at the present time, in the scientific literature. Several recent papers [17–21] have clearly shown that cosmic acceleration, as opposed to the cosmic coasting of $R_h = ct$ models, is not yet *proven*. These are not, of course, refutations of the existence of dark energy as it may be defined by general relativity. Rather, they are statistical analyses placing some doubt on dark energy *dominance* and thus cosmic *acceleration*. These papers are well worth reading.

4. Superiority of FSC compared to inflationary cosmology

As detailed in the recent FSC summary paper [10], there are at least 11 categories in which FSC appears to be superior to standard inflationary cosmology. What makes FSC so powerful in this regard is its ability to make very specific predictions for observations which can be used to falsify the theory if FSC is incorrect. To date, FSC as a global parameter observational predictor has not been falsified.

Standard inflationary cosmology, on the other hand, has largely been cobbled together from observations and would be difficult to falsify because it makes few, if any, falsifiable predictions. The reader should remember that the various theories of cosmic inflation contained ad hoc adjustments to accommodate observations [22, 23] and that the presumed "inflaton" energy field of inflation was invented before the actual cosmological vacuum energy now called dark energy was discovered approximately two decades later. It is notable that, rather than attempt to apply the newly discovered dark energy as a scalar quantity also at work in the early universe, standard inflationary cosmologists have generally assumed the dark energy field to be something entirely distinct from their theoretical inflaton energy field. There has also been an assumption that the post-inflationary energy density of the vacuum must have been a *constant* over the great span of cosmological time. And yet, the theoretical discrepancies created by this "cosmological constant problem" [24, 25] are considered by many to be the most embarrassing problem in all of physics. A discussion of this problem is included later.

What follows are several selected categories of particular importance from the FSC summary paper. The reader is encouraged to read this paper for the full discussion as to how FSC appears to be superior to standard inflationary cosmology, particularly in terms of falsifiability.

4.1 Cosmic dawn and the formation of the first quasars and galaxies

As noted in several recent papers [26, 27], standard inflationary cosmology cannot easily explain the surprisingly early formation of the first quasars and galaxies. As detailed in a recent FSC paper [28], temperature curve differences between the two models are such that cosmic dawn, at z redshifts of about 15–20, occurred in the FSC model much earlier than in standard inflationary cosmology. A comparison of the two temperature curves is shown below in **Figure 4**, with features of the standard inflationary model as illustrated in Bowman's recent paper [29].

The blue line is the radiation temperature (T_R) curve expected in standard inflationary cosmology, and the green line is the radiation temperature curve expected in FSC. The dashed red line represents the spin temperature (T_S), and the solid red line represents the baryonic gas temperature (T_G).

One should note how these cosmic times differ with respect to a given model's radiation temperature. Judging from these temperature curve differences, cosmic dawn in FSC would have been at about 20–50 million years after the Planck epoch as opposed to the standard inflationary cosmology cosmic dawn at about

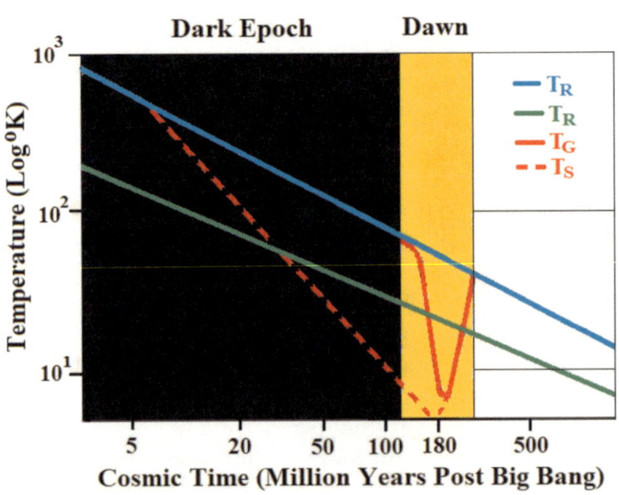

Figure 4.
Cosmic temperature vs. time in standard cosmology (blue) and FSC (green).

110–250 million years. Thus, FSC, by the relative flatness of its temperature curve, allows for considerably more time between the formation of the first stars and the formation of the first quasars and galaxies.

4.2 Predictions pertaining to primordial gravity waves

FSC is a steadily expanding cosmology model, which would not be expected to produce inflationary B-mode primordial gravity waves. There is nothing "explosive" about the FSC early universe in comparison with the standard inflationary early universe. Thus, FSC predicts that inflationary B-mode primordial gravity waves will never be detected. Such unequivocal detection of inflationary waves would falsify FSC. The continued failure to detect such waves, if the sensitivity of detection methods can be made sufficiently high, should be considered to strongly favor FSC over standard inflationary cosmology.

4.3 Predicting the magnitude of CMB temperature anisotropy

The angular power spectrum of the CMB clearly fits with a spatially flat universe. As noted following the BOOMERanG Collaboration report [30] of CMB anisotropy observations, their results are "closely fitting the theoretical predictions for a spatially flat cosmological model with an exactly scale invariant primordial power spectrum for the adiabatic growing mode" [31]. Furthermore, the COBE DMR experiment [32] measured a CMB RMS temperature variation of 18 micro-Kelvins. This translates to a dT/T anisotropy value of (0.000018)/2.725 equal to 0.66×10^{-5}. This measurement fits within the range of FSC temperature anisotropy predictions for the beginning and ending of the recombination/decoupling epoch [33]. This result clearly favors FSC.

4.4 Predicting the Hubble parameter value

In standard inflationary cosmology, the Hubble parameter value can only be determined by observation. That is to say that there is no theoretical ability within standard cosmology to derive a Hubble parameter value. The FSC model, on the other hand, *predicts* the current global H_o value to be 66.89 kilometers per second per megaparsec. This fits the 2018 Planck Collaboration [11] and 2018 DES [12] Hubble parameter values. Therefore, this category strongly favors FSC in comparison with standard inflationary cosmology.

4.5 Quantifiable entropy and the entropic arrow of time

One of the problems within the standard inflationary model is in quantifying cosmic entropy. Entropy is typically defined in terms of the total number of possible microstates and the probability of a given set of conditions with respect to that number of microstates. These values are impossible to quantify in an infinite-sized inflationary universe or multiverse. FSC, on the other hand, is a *finite* model with a spherical horizon surface area. And, since the Bekenstein-Hawking definition of black hole entropy applies to the FSC model, values for cosmic entropy can be calculated for any time, temperature, or radius of the FSC model. Thus, the "entropic arrow of time" is clearly defined and quantified in the FSC model. The quantifiable entropy of the FSC model allows for model correlations with cosmic entropy theories, such as those of Roger Penrose [34] and Erik Verlinde. Thus, the entropy rules of FSC potentially allow for falsifiability. This feature favors the FSC model, particularly with respect to Verlinde's "emergent gravity" theory (see below).

4.6 Clues to the nature of gravity, dark energy, and dark matter

The reader is referred to the recent FSC paper [35] with this title for an in-depth discussion of how cosmic entropy in the FSC model may provide tantalizing clues with respect to the fundamental nature of gravity. In short, the FSC model appears to be the cosmological model correlate to Verlinde's "emergent gravity" theory [36, 37]. Verlinde's landmark paper from 2011 provides strong theoretical support for gravity being an *emergent property* of cosmic entropy. The corresponding FSC paper makes a case for the correctness of Verlinde's theory. As discussed in the FSC paper, if gravity is an emergent property of cosmic entropy, then one might entertain the possibility that dark energy and dark matter could also be emergent properties of cosmic entropy. For instance, perhaps galactic and perigalactic features attributed to dark matter (such as platelike galactic rotation and gravitational lensing) could be an unexpected large-scale effect of the entropy of the known galactic baryonic matter. If this turns out to be the correct interpretation, then gravity, dark energy, and dark matter might be as difficult to define at the quantum level as "quantum consciousness" within two connected neurons.

The recent observations of Brouwer et al. [38] appear to be in support of Verlinde's "emergent gravity" theory as it pertains to dark matter. The discovery of quantum gravity, other than quantum gravity somehow connected to entropy at the quantum level, would falsify Verlinde's "emergent gravity" theory. At present, standard inflationary cosmology, by virtue of its inability to precisely define cosmic entropy, has no capacity to incorporate Verlinde's theory. This appears to favor FSC, particularly in light of the above-mentioned recent observational findings.

4.7 The cosmological constant problem

The "cosmological constant problem" is a long-standing problem in theoretical physics. It underscores standard cosmology's inability to unify general relativity with quantum field theory (QFT). Excellent expositions on this subject have been provided by Weinberg [24] and Carroll [25]. QFT theorists calculate a cosmological constant value which differs from observational measurements of the vacuum energy density by a magnitude of approximately 10^{121}! Suffice it to say, this discrepancy is so large that it is often referred to as the most embarrassing problem in all of theoretical physics.

In standard inflationary cosmology, it has been assumed that the post-inflationary energy density of the cosmic vacuum must be constant, rather than scalar, over the remainder of cosmic time. However, general relativity does indeed allow for the vacuum energy density to be a dynamic scalar over time, so long as $\Lambda = 3H_t^2/c^2$. Cosmological models incorporating scaling vacuum energy density are called "quintessence" models. FSC is one such model. In FSC, the vacuum energy density scales downward by 121.26 logs of 10 over the cosmic time interval since the Planck epoch. Perhaps of even greater interest is that the Bekenstein-Hawking cosmic entropy value scales upward in direct proportion to the expanding surface area of the cosmic horizon. If one were to count the current number of Planck radius microstates within the FSC horizon, the model indicates this entropy number to be $10^{121.26}$. Thus, by its implication of a possible relationship between vacuum energy (i.e., dark energy) and total cosmic entropy (as discussed in Section 4.6), FSC also offers a possible explanation for the magnitude difference between the Planck epoch vacuum energy density calculated by QFT theorists and today's observed vacuum energy density of approximately 10^{-9} J.m^{-3}. Since the FSC model stipulates these values and standard inflationary cosmology has no basis for deriving them, the FSC model appears to be superior with respect to potentially resolving the cosmological constant problem.

4.8 Dark matter and dark energy quantitation

As reported by the Planck Collaboration, the ratio of dark matter to visible (baryonic) matter is observed to be approximately 5.5 parts dark matter to 1 part visible matter. However, there are already significant differences observed between the dark matter-to-visible matter ratios in the galaxies quite near to us (essentially co-movers) and the above dark matter-to-visible matter ratio determined from Planck CMB observations. *Perhaps this ratio is scalar over the great span of cosmic time. If the co-mover ratio is ultimately found to be approximately 9.2, as predicted by FSC, one can then conclude that total matter energy density at present is equal in absolute magnitude to dark energy density*. This equality of opposite sign energy densities is what one would expect for a spatially flat universe. Otherwise, if one energy density dominated the other, there should be detectable global spatial curvature corresponding to the dominating energy density. One could, in fact, make a strong case that the spatial flatness of the CMB proves the equality of total matter and dark energy densities at the time of the recombination/decoupling epoch. This should nullify any Planck Collaboration conclusions (such as dark energy dominance) which are obviously contrary to their own observations of spatial flatness.

Despite the fact that FSC and standard inflationary cosmology differ somewhat with respect to the percentages of total matter vs. dark energy predicted for the co-moving universe, there is one thing about this energy density partition on which everyone agrees: it is truly remarkable that total matter energy density and dark energy density are *of the same order of magnitude at the present time*. As physicist I. I. Rabi once famously remarked, "Who ordered that?!" This is often referred to as the cosmological "coincidence problem." Standard cosmology simply accepts this coincidence problem with no further explanation or rationale. However, FSC *stipulates* perpetual equality of absolute magnitude of these two energy densities as a requirement for a perpetually spatially flat universe. One can consider this expectation of energy density equality to be a falsifiable FSC prediction with respect to future measurements of total matter energy density in comparison with dark energy density. *An in-depth statistical analysis of approximate co-movers with the Milky Way should give us a better idea of the dark matter-to-visible matter ratio in the current epoch*.

With respect to standard cosmology's current belief in cosmic acceleration due to dark energy, the reader is referred to the references [17] thru [21] mentioned earlier. Cosmic acceleration is clearly not proven at the present time, despite the indisputable presence of dark energy as definable within general relativity. There are relative differences in luminosity distance and angular diameter distance formulae in standard inflationary cosmology and $R_h = ct$ modified Milne-type models (like FSC). Two comparative graphs from FSC reference [39] are shown in **Figures 5** and **6**.

The significance of the relative luminosity distance and relative angular diameter distance comparisons between these two competing models is paramount. An observer of distant Type Ia supernovae expects particular luminosity distances and angular diameter distances to correspond with particular redshifts. If, instead, he or she observes greater-than-expected luminosity distances (i.e., unexpected "dimming" of the supernovae) or greater-than-expected angular diameter distances, this can easily be misinterpreted by a standard inflationary model proponent as indicative of cosmic acceleration. However, *entirely predictable supernova luminosity distances within a realistic Milne-type universe containing matter, as opposed to a standard model universe, could be one possible explanation for the Type Ia supernovae observations since 1998. Obviously, cosmic acceleration would not then be required to explain these observations*. This possibility, combined with the standard model tension problem presented above (i.e., spatial flatness and dark energy dominance cannot *both* be

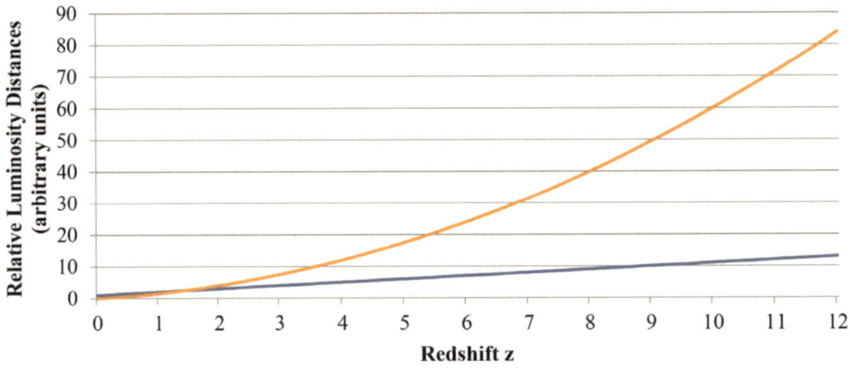

Figure 5.
Relative luminosity distances vs. redshift z for standard (blue) and Milne (red).

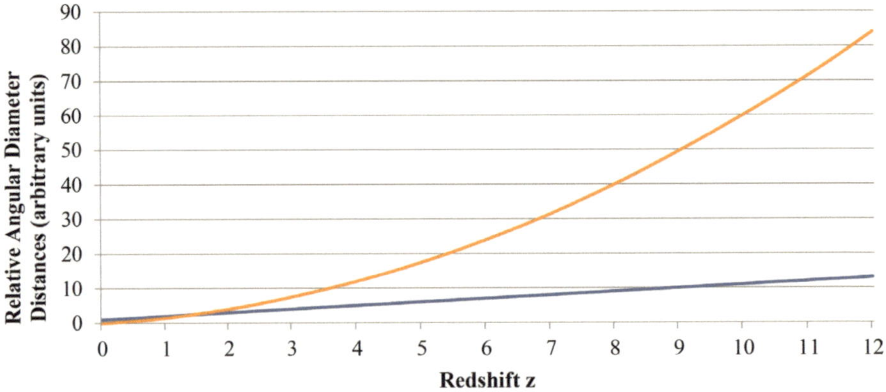

Figure 6.
Relative angular diameter distances vs. z for standard (blue) and Milne (red).

true at the same time), and the FSC *stipulation* of what standard model proponents refer to as the "coincidence problem," strongly favors FSC with respect to its predictions concerning dark matter and dark energy quantitation.

4.9 Requirements for new physics

Cosmic inflation theory was invented before the spatial-flattening effects (on positively curved space-time) of cosmic vacuum energy (dark energy) were discovered in 1998 [40–42]. Guth [43] and others [44, 45] believed at the time of its invention that a special energy field with inflating features (called by Guth the "inflaton") was required within the initial 10^{-32} s of universal expansion. It was believed that this energy field was necessary in order to flatten out a presumed highly curved space-time during and immediately following the inception of expansion. Thus, inflation appeared to be a clever solution to the cosmological "flatness problem," as well as the cosmological "horizon problem." The latter problem was presumed at the time to exist because most cosmologists believed, without any real evidence, that the universe is infinite and thus otherwise difficult to explain in terms of its remarkable homogeneity in all observational directions.

For reasons mentioned near the end of the "Introduction and Background" section, FSC solves these cosmological problems without requiring an inflationary epoch. In contrast to inflationary models, in which the total cosmic matter generation is exclusively limited to within a tiny fraction of a second of the Big Bang, *the FSC model is a perpetual matter-generating model* with some similarities to the model presented in the 2019 publication entitled “A Perpetual Mass-Generating Planckian Universe” by Sapar [46]. This concept of perpetual matter generation has a long tradition going back at least to Hoyle, although Hoyle’s particular matter-generating theory was falsified by the discovery of the cosmic microwave background in the 1960s. Here it is important to recognize that the mystery of matter generation is inherent in *all* cosmology models. FSC simply models perpetual matter generation, while inflationary models imply, without any real evidence, that all universal matter was nearly instantaneously created.

This author speculates that the negative energy (i.e., gravitationally repelling) vacuum may be continually diluted of its original highly concentrated Planckian energy during cosmic expansion and that gravitationally attracting positive energy in the form of matter is continually created as an offset. This would be in keeping with the spatial curvature rules of general relativity. One should remember that, according to general relativity, a flat space-time is flat precisely because it contains net zero total energy. Furthermore, a globally and perpetually spatially flat universe which begins from a net zero total energy state (Guth’s “free lunch” idea) would presumably maintain net zero total energy throughout its expansion. Otherwise, a fully self-contained universe, such as a FSC universe, would violate conservation of energy.

Despite the ongoing mystery of matter generation in all cosmology models, for the arguments made above, and for the perpetual matter generation rationale offered in Dr. Sapar’s paper, this category appears to favor FSC in comparison with standard inflationary cosmology.

5. Summary and conclusions

This chapter has introduced the reader to the heuristic FSC cosmology model. Like all useful heuristics, FSC provides a means for accurately calculating a variety of parameters. The founding principle for the construction of this model is Hawking’s singularity theorem. Accordingly, all assumptions of this model are intrinsically linked to Hawking’s theorem and its implications with respect to widely accepted time-symmetric properties of general relativity. Black holes and black hole-like objects are now known to exist. Furthermore, we know that such objects range over a remarkably wide, fractal-like scale. Our universe may simply be the largest of these objects which can be observed, albeit from the inside!

Beginning with Penrose and Hawking, the black hole-like properties of the universe have continued to fascinate and surprise us. Our current golden age of astrophysical observations and new theories certainly promises even more surprises ahead.

Dedications and acknowledgments

This paper is dedicated to Stephen Hawking and Roger Penrose for their groundbreaking work on black holes and their possible application to cosmology. Dr. Tatum sincerely thanks U.V.S. Seshavatharam for his co-authorship of the seminal FSC papers and some of the more recent FSC publications. He also thanks Dr. Rudolph Schild of the Harvard-Smithsonian Center for Astrophysics for his past support and encouragement.

Author details

Eugene Terry Tatum
Independent Researcher, Bowling Green, Kentucky, USA

*Address all correspondence to: ett@twc.com

References

[1] Penrose R. Gravitational collapse and space-time singularities. Physical Review Letters. 1965;**14**(3):57-59

[2] Hawking S. Properties of expanding universes. 1966. Cambridge University Library; PhD. 5437. Available from: https://cudl.lib.cam.ac.uk/view/MS-PHD-05437/134

[3] Hawking S, Penrose R. The singularities of gravitational collapse and cosmology. Proceeding of the Royal Society of London. 1970;**314**:529-548

[4] Tatum ET, Seshavatharam UVS, Lakshminarayana S. The basics of flat space cosmology. International Journal of Astronomy and Astrophysics. 2015;**5**: 116-124. DOI: 10.4236/ijaa.2015.52015

[5] Tatum ET, Seshavatharam UVS, Lakshminarayana S. Thermal radiation redshift in flat space cosmology. Journal of Applied Physical Science International. 2015;**4**(1):18-26

[6] Tatum ET, Seshavatharam UVS, Lakshminarayana S. Flat space cosmology as an alternative to LCDM cosmology. Frontiers of Astronomy, Astrophysics and Cosmology. 2015;**1**(2): 98-104. Available from: http://pubs.sciepub.com/faac/1/2/3. DOI: 10.12691/faac-1-2-3

[7] Bekenstein JD. Generalized second law of thermodynamics in black hole physics. Physics Review D. 1974;**9**: 3292-3300. DOI: 10.1103/PhysRevD.9.3292

[8] Hawking S. Black holes and thermodynamics. Physical Review D. 1976;**13**(2):191-197

[9] Borde A et al. Inflationary spacetimes are not past-complete. Physical Review Letters. 2003;**90**(15):151301. arXiv:gr-qc/0110012

[10] Tatum ET. Why flat space cosmology is superior to standard inflationary cosmology. Journal of Modern Physics. 2018;**9**:1867-1882. DOI: 10.4236/jmp.2018.910118

[11] Aghanim N, et al. Planck 2018 Results VI. Cosmological Parameters. 2018. Available from: http://arXiv: 1807.06209v1

[12] Macaulay E et al. First cosmological results using type Ia supernovae from the dark energy survey: Measurement of the hubble constant. 2018. arXiv: 1811.02376v1

[13] Inoue AK. Amount of intergalactic dust: Constraints from distant supernovae and the thermal history of the intergalactic medium. Monthly Notices of the Royal Astronomical Society. 2004;**350**(2):729-744. DOI: 10.1111/j.1365-2966.2004.07686.x

[14] Tatum ET, Seshavatharam UVS. Flat space cosmology as a model of light speed cosmic expansion—Implications for the vacuum energy density. Journal of Modern Physics. 2018;**9**:2008-2020. DOI: 10.4236/jmp.2018.910126

[15] Suzuki N et.al. The hubble space telescope cluster supernovae survey: V. Improving the dark energy constraints above Z > 1 and building an early-type-hosted supernova sample. 2011. arXiv.org/abs/1105.3470

[16] van Dokkum P et al. A galaxy lacking dark matter. Nature. 2018;**555**: 629-632. DOI: 10.1038/nature25767

[17] Melia F. Fitting the union 2.1 SN sample with the R_h = ct universe. Astronomical Journal. 2012;**144**:110-135. arXiv:1206.6289 [astro-ph.CO]

[18] Nielsen JT et al. Marginal evidence for cosmic acceleration from type Ia

supernovae. Scientific Reports. 2015;**6**: 35596. DOI: 10.1038/srep35596. arXiv: 1506.01354v1

[19] Wei J-J et al. A comparative analysis of the supernova legacy survey sample with ΛCDM and the R_h = ct universe. Astronomical Journal. 2015;**149**:102

[20] Tutusaus I et al. Is cosmic acceleration proven by local cosmological probes? Astronomy & Astrophysics. 2017;**602**:A73. arXiv: 1706.05036v1 [astro-ph.CO]

[21] Dam LH et al. Apparent cosmic acceleration from type Ia supernovae. Monthly Notices of the Royal Astronomical Society. 2017;**472**. arXiv: 1706.07236v2 [astro-ph.CO]

[22] Guth AH. The Inflationary Universe. New York, US: Basic Books; 1997. p. 238

[23] Steinhardt PJ. The inflation debate: Is the theory at the heart of modern cosmology deeply flawed? Scientific American. 2011;**304**(4):18-25

[24] Weinberg S. The cosmological constant problem. Reviews of Modern Physics. 1989;**61**:1-23

[25] Carroll S. The cosmological constant. Living Reviews in Relativity. 2001;**4**:5-56

[26] Hashimoto T et al. The onset of star formation 250 million years after the big bang. 2018. arXiv.1805.05966v1 [astro-ph.GA]. [Accessed: 15 May 2018]

[27] Natarajan P. The first monster black holes. Scientific American. 2018;**318**(2): 24-29

[28] Tatum ET, Seshavatharam UVS. Temperature scaling in flat space cosmology in comparison to standard cosmology. Journal of Modern Physics. 2018;**9**:1404-1414. DOI: 10.4236/ jmp.2018.97085

[29] Bowman JD. An absorption profile centered at 78 Megahertz in the sky-ave raged spectrum. Nature. 2018;**555**:67-70. DOI: 10.1038/nature25792

[30] de Bernardis P et al. A flat universe from high-resolution maps of the cosmic microwave background radiation. 2000. arXiv:astro-ph/ 0004404v1. DOI: 10.1038/35010035

[31] Bucher M. Physics of the cosmic microwave background anisotropy. 2015. arXiv:1501.04288v1

[32] Wright EL et al. Angular power spectrum of the cosmic microwave background anisotropy seen by the COBE DMR. Astrophysical Journal. 1996;**464**:L21-L24. DOI: 10.1086/ 310073

[33] Tatum ET. How the CMB anisotropy pattern could be a map of gravitational entropy. Journal of Modern Physics. 2018;**9**:1484-1490. DOI: 10.4236/jmp.2018.98092

[34] Penrose R. Fashion Faith and Fantasy in the New Physics of the Universe. Princeton, US: Princeton University Press; 2016

[35] Tatum ET, Seshavatharam UVS. Clues to the fundamental nature of gravity, dark energy and dark matter. Journal of Modern Physics. 2018;**9**: 1469-1483. DOI: 10.4236/jmp. 2018.98091

[36] Verlinde E. On the origin of gravity and the Laws of Newton. Journal of High Energy Physics. 2011;**4**:29-55. arXiv:1001.0785v1 [hep-th]. DOI: 10.1007/JHEP04(2001)029

[37] Verlinde E. Emergent gravity and the dark universe. 2016. arXiv: 1611.02269v2 [hep-th]

[38] Brouwer MM et al. First test of Verlinde's theory of emergent gravity using weak gravitational lensing

measurements. Monthly Notices of the Royal Astronomical Society. 2016;**466**: 1-14. arXiv:1612.03034v2 [astro-ph.CO]

[39] Tatum ET, Seshavatharam UVS. How a realistic linear R_h = ct model of cosmology could present the illusion of late cosmic acceleration. Journal of Modern Physics. 2018;**9**:1397-1403. DOI: 10.4236/jmp.2018.97084

[40] Perlmutter S et al. The supernova cosmology project, measurements of omega and lambda from 42 high-redshift supernovae. Astrophysical Journal. 1999;**517**:565-586. [astro-ph/9812133]

[41] Schmidt B et al. The high-Z supernova search: Measuring cosmic deceleration and global curvature of the universe using type Ia supernovae. Astrophysical Journal. 1998;**507**:46-63

[42] Riess AG et al. Observational evidence from supernovae for an accelerating universe and a cosmological constant. Astronomical Journal. 1998; **116**(3):1009-1038

[43] Guth AH. Inflationary universe: A possible solution to the horizon and flatness problems. Physics Review D. 1981;**23**:347

[44] Albrecht A, Steinhardt PJ. Cosmology for grand unified theories with radiatively induced symmetry breaking. Physical Review Letters. 1982; **48**:1220-1223

[45] Linde AD. A new inflationary universe scenario: A possible solution of the horizon, flatness, homogeneity, isotropy, and primordial monopole problems. Physics Letters. 1982;**108B**: 389-392

[46] Sapar A. A perpetually mass-generating planckian universe. Proceedings of the Estonian Academy of Sciences. 2019;**68**(1):1-12. DOI: 10.3176/proc.2019.1.01

Chapter 3

Thermal Stability Criteria of a Generic Quantum Black Hole

Aloke Kumar Sinha

Abstract

Thermodynamics of black holes were studied by Hawking, Bekenstein et al., considering black holes as classical spacetimes possessing a singular region hidden behind an event horizon. In this chapter, in contrast, we treat black hole from the perspective of a generic theory of quantum gravity, using certain assumptions which are consistent with loop quantum gravity (LQG). Using these assumptions and basic tenets of equilibrium statistical mechanics, we have derived criteria for thermal stability of black holes in any spacetime dimension with arbitrary number of charges ('hairs'), irrespective of whether classical or quantum. The derivation of these thermal stability criteria makes no explicit use of classical spacetime geometry at all. The only assumption is that the mass of the black hole is a function of its horizon area and all the 'hairs' (i.e. charge, angular momentum, any other types of hairs). We get a series of inequalities between derivatives of the mass function with respect to the area and other 'hairs' as the thermal stability criteria. These criteria are then tested in detail against various types of black holes in various dimensions. This permits us to predict the region of the parameter space of a given black hole in which it may be stable under Hawking radiation.

Keywords: black hole thermodynamics, thermal stability, saddle-point approximation, quantum gravity, multicharged black hole
PAC numbers: 04.70.-s, 04.70.Dy

1. Introduction

Semiclassical analysis has made the claim that non-extremal, asymptotically flat black holes are thermally unstable due to decay under Hawking radiation. Their instability is allegedly due to negativity of their specific heat [1, 2], as deduced from semiclassical mnemonics based on the classical metric. These black holes become hotter and hotter as they lose mass. This is a complete thermal runaway process. Note, however, that semiclassical analysis depends explicitly on the classical black hole metric and, as such, is inherently a 'case-by-case' analysis. This limitation implies that general results about thermal stability of black holes under Hawking decay cannot be obtained from such an analysis. For some asymptotically flat general relativistic black holes, semiclassical analysis has yielded the understanding that their specific heat, defined semiclassically from their metric, is negative, and hence the black holes must be thermally unstable under Hawking decay. However, there is little to glean from this approach which holds in general.

This interesting fact has motivated the study of thermal stability of black holes, from a perspective that is inspired by a definite proposal for *quantum* spacetime (like loop quantum gravity (LQG) [3, 4]) rather than on semiclassical assumptions. In the vicinity of a black hole horizon, gravity is very strong. So, a nonperturbative quantum theory of gravity is required to describe black holes from a quantum perspective. LQG is one of the promising candidates having this feature. A consistent understanding of the issue of *quantum* black hole entropy has been obtained through LQG [5, 6], where not only has the Bekenstein-Hawking area law been retrieved for macroscopic (astrophysical) black holes, but a whole slew of corrections to it, due to quantum spacetime fluctuations that have been derived as well [7, 8], with the leading correction being logarithmic in area with the coefficient $-3/2$. LQG plays only a motivational role in our work. Many of the assumptions, actually are made independently of LQG, are justified on the ground that LQG might provide situations where these assumptions are valid.

The implications of this quantum perspective on the thermal stability of black holes from decay due to Hawking radiation have therefore been an important aspect of black hole thermodynamics beyond semiclassical analysis and also somewhat beyond the strict equilibrium configurations that isolated horizons represent. Classically a black hole in general relativity is characterized by its mass (M), charge (Q) and angular momentum (J). Intuitively, therefore, we expect that thermal behaviour of black holes will depend on all of these parameters. For a given classical metric characterizing a black hole, the mass can be derived explicitly to be a function of the charge and angular momentum. However, the quantum spacetime perspective frees us from having to use classical formulae for this functional dependence of the mass. Instead, the assumption is simply this: the mass is a function of the horizon *area*, along with the charge and angular momentum.

The simplest case of vanishing charge and angular momentum has been investigated longer than a decade ago [9–11]. The obtained condition for thermal stability exactly matches with the condition, derived from semiclassical analysis. That condition has been derived from positivity of specific heat. This exact matching happens as the black holes have neither rotation nor charge. We are going to establish in this chapter that even if a black hole has at least one of those, the conditions for thermal stability are more elaborate. This is already obvious when one considers charged black holes ([12]). Therefore the conditions start to differ from classical ones. This is due to the fact that black holes are treated quantum mechanically. The earlier work has been generalized, via the idea of *thermal* holography ([13, 14]) and the saddle-point approximation to evaluate the canonical partition function corresponding to the horizon, retaining Gaussian thermal fluctuations. The consequence is a general criterion of thermal stability as an inequality connecting area derivatives of the mass and the microcanonical entropy. This inequality is nontrivial when the microcanonical entropy has corrections (of a particular algebraic sign) beyond the area law, as is the case for the loop quantum gravity calculation of the microcanonical entropy [15]. The generalized stability criterion indeed 'predicts' the thermal *instability* of asymptotically flat Reissner-Nordstrom black holes contrasted with the thermal *stability* of anti-de Sitter Reissner-Nordstrom black holes (for a range of cosmological constants).

In this chapter, this approach is generalized to quantum black holes carrying *both* charge and angular momentum. The inclusion of rotation poses challenges in the LQG formulation [16–19] of isolated horizons. However, the general understanding of nonradiant rotating isolated horizons has parallels in these assays. We do not review this body of work, but realize that the thermal stability behaviour of rotating radiant black holes may be *qualitatively different* from that of the nonrotating ones.

We have calculated the partition function for rotating charged black hole. Thereafter we have got several inequalities as criteria for thermal stability of such black hole. We interpret these criteria and show how they are related to various thermodynamical quantities. We also show how the stability criteria for nonrotating and neutral black holes can be derived from these seven conditions in appropriate limit.

Beyond the standard general relativity theory corresponding to $(3+1)$ dimensional spacetime, higher-dimensional theories of gravity are currently under extensive scrutiny. Consequently black hole solutions are also being considered in those theories [20]. Such black holes have additional charges (hairs) beyond the traditional hairs—electric charge and angular momentum[1]. A black hole can be completely designated by its charge (Q), mass or area (A) and angular momentum (J). Quantum mechanically, a black hole can have many extra hairs, i.e. many charges, which contribute to its mass [22–28]. Higher-dimensional black holes too have many new charges which contribute to its mass [29]. We also consider all such hairs of black holes, together. We generalize the analysis of thermal stability of $(3+1)$ dimensional charged rotating black holes, for black holes with arbitrary number of hairs in any spacetime dimension. We find that the process of generalization is reasonably straightforward, except for calculational complications.

2. Quantum algebra and black hole spectrum

Like for all quantum systems, an operator algebra of fundamental observables is required to have a proper quantum description of black holes. Classically, generic black holes are represented by four parameters (M, Q, J, A), with three of them being independent. So, this naturally raises the question of the choice of the trivalent subset of classical variables that are promoted to quantum operators. Hence that three will be the fundamental observables of the quantum theory of black hole, and the remaining variable would correspond to a secondary observable.

Now, it is not possible to have a rotating, charged black hole without any mass, i.e. $M = 0$ with $Q, J \neq 0$. In fact an uncharged, nonrotating black hole does have mass, i.e. $M \neq 0$ with $Q = J = 0$. Therefore, charge and angular momentum are additional structures that can be imposed on a black hole. Hence they are preferably fundamental observables in a quantum theory of black hole.

We can choose any one between area (A) and mass (M) as the third fundamental observable. We choose area (A) as the third fundamental observable. So, mass (M) becomes the secondary observable, i.e. $M = M(A, Q, J)$. So, the algebraic approach of black hole quantization gives $\widehat{Q}, \widehat{J}, \widehat{A}$ as quantum operators of fundamental observables and $\widehat{M}\left(\widehat{H}_b\right)$ as quantum operator of secondary observable. All these correspond to the isolated horizon of a black hole.

It is physically obvious that both area and charge should be invariant under SO (3) rotations and area should also be invariant under U(1) gauge transformation. Now, angular momentum is the generator for rotation ($SO(3)$ Group), and charge is the generator of the U(1) global gauge group. These give

$$\left[\widehat{A}, \widehat{J}\right] = \left[\widehat{A},\ Q\right] = \left[\widehat{Q}, \widehat{J}\right] = 0 \tag{1}$$

---

[1] 'No hair' theorem [21–29] for black holes states that black hole cannot have any hair classically

Since $\widehat{M}\left(\widehat{H}_b\right)$ is a quantum operator of secondary observable $(M(A,J,Q))$, Eq. (1) can be extended as

$$\left[\widehat{A},\widehat{J}\right] = \left[\widehat{A},\widehat{Q}\right] = \left[\widehat{A},\widehat{M}\right] = \left[\widehat{Q},\widehat{J}\right] = \left[\widehat{M},\widehat{Q}\right] = \left[\widehat{J},\widehat{M}\right] = 0 \tag{2}$$

Hence, $\widehat{A},\widehat{J},\widehat{Q}$ can have simultaneous eigenstate. This implies the fact that a black hole can have definite values of charge, area and angular momentum up to thermodynamical and quantum fluctuation. In fact these eigenvalues of $\widehat{A},\widehat{J},\widehat{Q}$ are precisely the values that are used in the classical metric of a black hole to express mass (M) as a function of them.

3. Thermal holography

Quantum black holes associated with an ambient thermal reservoir have been considered in the past [9–11, 13, 30]. In this approach key results of LQG like the discrete spectrum of the area operator [3, 4] have been used, and the main assumption was that the thermal equilibrium configuration is indeed an isolated horizon (IH) whose microcanonical entropy, including quantum spacetime fluctuations, has already been computed via LQG. The idea was to study the interplay between thermal and quantum fluctuations, and a criterion for thermal stability of such horizons has been obtained [11, 13, 14], using a 'thermal holographic' description involving a canonical ensemble and incorporating Gaussian thermal fluctuations. The generalization to horizons carrying charge has also been attempted, using a grand canonical ensemble, even though a somewhat ad hoc mass spectrum has been assumed [10].

Here, we attempt to generalize the thermal holography for nonrotating electrically charged quantum radiant horizons discussed in [12], to the situation when the horizon has both charge and angular momentum, without any ad hoc assumptions on the mass spectrum. Such a generalization completes the task set out in [9, 13] to include charge and angular momentum simultaneously in consideration of thermal stability of the horizon under Hawking radiation. A comparison with semiclassical thermal stability analysis of black holes [31] is made wherever possible.

3.1 Mass associated with horizon

Black holes at equilibrium are represented by isolated horizons, which are internal boundaries of spacetime. Hamiltonian evolution of this spacetime gives the first law associated with isolated horizon (b), assumed to be a null hypersurface with the properties of a 'one-way membrane' [16, 32]. The law is given as

$$\delta E_h^t = \frac{\kappa^t}{8\pi}\delta A_h + \Phi^t \delta Q_h + \Omega^t \delta J_h \tag{3}$$

where E_h^t is the energy function associated with the horizon; κ^t, Φ^t and Ω^t are, respectively, the surface gravity, electric potential and angular velocity of the horizon; and Q_h, A_h and J_h are, respectively, the charge, area and angular momentum of the horizon. The label 't' denotes the particular time evolution field (t^μ) associated with the spatial hypersurface chosen. κ^t, Φ^t and Ω^t are defined for this particular choice of time evolution vector field t^μ. The family of time evolution vector fields $[t^\mu]$ satisfying such first laws on the horizon are the permissible time evolution vector fields. These evolution vector fields also need to satisfy other boundary conditions. Each of these time evolution vector fields associates an energy function with the horizon which is a function of area, charge and angular momentum, i.e. E_h^t is a function of A_h, Q_h and J_h.

The advantage of the isolated (and also the radiant or *dynamical*) horizon description is that one can associate with it a mass M_h^t, related to the ADM energy of the spacetime through the relation

$$E_{ADM}^t = M_h^t + H_{rad}^t \tag{4}$$

where H_{rad}^t is the Hamiltonian associated with spacetime between the horizon and asymptopia. It is the Hamiltonian of the covariant phase space, which is the space of various classes of solutions of the Einstein equations admitting internal boundaries. For stationary spacetimes the global time-like Killing field (ξ^μ) is the time evolution vector field. There is nothing between the internal boundary and asymptopia for stationary spacetimes; hence $H_{rad}^\xi = 0$. Actually, H_{rad}^ξ generates evolution along ξ^μ. So, for the stationary black hole, H_{rad}^ξ must vanish as a first-class constraint on the phase space [16, 33]. This gives $M_h^\xi = E_{ADM}^\xi$. This implies that for stationary black hole spacetimes, the ADM mass equals the energy of the black hole. Hence it is legitimate to identify E_h^ξ with the horizon mass M_h in the stationary case. The difference for an arbitrary nonstationary case is that $H_{rad}^t \neq 0$. Thus it can be called as the mass associated with the isolated horizon. So, an isolated horizon does not require stationarity, and therefore admits $H_{rad}^t \neq 0$, and hence admits a mass defined *locally* on the horizon, since the theory is topological and insensitive to small metric deformations.

Clearly, the horizon mass is *not* affected by boundary conditions at asymptopia. It is defined *locally* on the horizon without knowing the asymptotic structure at all. The asymptotic conditions only modify the energy associated with asymptopia and the bulk equation of motion (Einstein equations) [16, 34]. This Hamiltonian framework above is also applicable for both asymptotically flat and AdS spacetimes.

3.2 Quantum geometry

The boundary conditions of a classical spacetime with boundary determine the boundary degrees of freedom and their dynamics. For a quantum spacetime, fluctuations of the boundary degrees of freedom have a 'life' of their own [5, 6]. Therefore the Hilbert space of a quantum spacetime with boundary has the tensor product structure $\mathcal{H} = \mathcal{H}_b \otimes \mathcal{H}_v$, where the subscript b (v) denotes the boundary (bulk) component.

Thus, a generic quantum state $(|\Psi\rangle)$ can be expanded as

$$|\Psi\rangle = \sum_{b,v} C_{b,v} |\chi_b\rangle \otimes |\psi_v\rangle \tag{5}$$

where $|\chi_b\rangle$ is the boundary part of the full quantum state and $|\psi_v\rangle$ denotes the bulk component of the full quantum state.

The total Hamiltonian operator $(\widehat{H})$ acting on the generic state $(|\Psi\rangle)$ is given as

$$\widehat{H}|\Psi\rangle = \left(\widehat{H_b} \otimes I_v + I_b \otimes \widehat{H_v}\right)|\Psi\rangle \tag{6}$$

where, respectively, $I_b(I_v)$ is the identity operator on $\mathcal{H}_b(\mathcal{H}_v)$ and $\widehat{H_b}\left(\widehat{H_v}\right)$ is the Hamiltonian operator on $\mathcal{H}_b(\mathcal{H}_v)$.

In the presence of electric charge and rotation, $|\psi_v\rangle$ will be the composite bulk state. Hence, these bulk states are annihilated by the *full* bulk Hamiltonian, i.e.

$$\widehat{H_v}|\psi_v\rangle = 0 \tag{7}$$

This is the quantum version of the classical Hamiltonian constraint [4].

The charge operator ($\widehat{Q}$) for a black hole is defined as

$$\widehat{Q}|\Psi\rangle = \left(\widehat{Q}_b \otimes \widehat{I}_v + \widehat{I}_b \otimes \widehat{Q}_v\right)|\Psi\rangle \tag{8}$$

where, respectively, $\widehat{Q}_b$ and $\widehat{Q}_v$ are corresponding charge operators for the boundary states ($|\chi_b\rangle$) and the bulk states ($|\psi_v\rangle$).

Classically, the charge of a black hole is defined on the horizon, i.e. the internal boundary of the spacetime (e.g. one can see how charge can be properly defined for spacetimes admitting internal boundaries in Einstein-Maxwell or Einstein-Yang-Mills theories in [32]). There is *no* charge associated with the bulk black hole spacetime, i.e. $Q_v \approx 0$, which is basically the Gauss law constraint for electrodynamics. Hence, its quantum version is of the form:

$$\widehat{Q}_v|\psi_v\rangle = 0 \tag{9}$$

Like the charge operator, angular momentum operator ($\widehat{J}$) of a black hole can be defined as

$$\widehat{J}|\Psi\rangle = \left(\widehat{J}_b \otimes \widehat{I}_v + \widehat{I}_b \otimes \widehat{J}_v\right)|\Psi\rangle \tag{10}$$

where, respectively, $\widehat{J}_b$ and $\widehat{J}_v$ are corresponding angular momentum operators for the boundary states ($|\chi_b\rangle$) and the bulk states ($|\psi_v\rangle$).

A generic quantum bulk Hilbert space is invariant under local spacetime rotations, as a part of local Lorentz invariance. Angular momentum is the generator of spacetime rotation. Therefore it implies that bulk states are annihilated by angular momentum operator, i.e.

$$\widehat{J}_v|\psi_v\rangle = 0 \tag{11}$$

Hence Eqs. (7), (9) and (11) together give

$$\left[\widehat{H_v} - \beta\Phi\widehat{Q_v} - \beta\Omega\widehat{J_v}\right]|\psi_v\rangle = 0 \tag{12}$$

where β, Φ and Ω can be any function. But we will see that those will correspond to inverse temperature, electric potential and angular velocity, respectively, in afterwards.

3.3 Grand canonical partition function

We now consider a grand canonical ensemble of quantum spacetimes with horizons as boundaries, in contact with a heat bath, at some (inverse) temperature β. We will assume that this grand canonical ensemble of massive rotating charged black holes can exchange energy, angular momentum and charge with the heat bath. Therefore the grand canonical partition function is then given as

$$Z_G = Tr\left(exp\left(-\beta\widehat{H} + \beta\Phi\widehat{Q} + \beta\Omega\widehat{J}\right)\right) \tag{13}$$

over all states. $\widehat{Q}$ is the charge operator for the black hole and Φ is the corresponding electrostatic potential. Similarly $\widehat{J}$ is the angular momentum operator for the black hole, and Ω is the corresponding angular velocity.

The above definition, together with Eqs. (5), (6), (8), (10) and (12), yields

$$\begin{aligned}
Z_G &= \sum_{b,v} |C_{b,v}|^2 \left\langle \psi_v \right| \otimes \left\langle \chi_b \right| exp\left(-\beta\widehat{H} + \beta\Phi\widehat{Q} + \beta\Omega\widehat{J}\right) \left|\chi_b\right\rangle \otimes \left|\psi_v\right\rangle \\
&= \sum_{b,v} |C_{b,v}|^2 \left\langle \psi_v \right| exp\left(-\beta\widehat{H}_v + \beta\Phi\widehat{Q}_v + \beta\Omega\widehat{J}_v\right) \left|\psi_v\right\rangle \left\langle \chi_b \right| exp\left(-\beta\widehat{H}_b + \beta\Phi\widehat{Q}_b + \beta\Omega\widehat{J}_b\right) \left|\chi_b\right\rangle \\
&= \sum_{b,v} |C_{b,v}|^2 \langle \psi_v | \psi_v \rangle \left\langle \chi_b \right| exp\left(-\beta\widehat{H}_b + \beta\Phi\widehat{Q}_b + \beta\Omega\widehat{J}_b\right) \left|\chi_b\right\rangle \\
&= \sum_b \left(\sum_v |C_{b,v}|^2 \langle \psi_v | \psi_v \rangle \right) \langle \chi_b | exp(-\beta\widehat{H}_b + \beta\Phi\widehat{Q}_b + \beta\Omega\widehat{J}_b | \chi_b \rangle \\
&= \sum_b |C_b|^2 \left\langle \chi_b \right| exp\left(-\beta\widehat{H}_b + \beta\Phi\widehat{Q}_b + \beta\Omega\widehat{J}_b\right) \left|\chi_b\right\rangle
\end{aligned} \tag{14}$$

assuming that the boundary states can be normalized through the squared norm $\sum_v |c_{vb}|^2 \langle \psi_v | \psi_v \rangle = |C_b|^2$. This is analogous to the canonical ensemble scenario described in [13].

The partition function thus turns out to be completely determined by the boundary states (Z_{Gb}), i.e.

$$Z_G = Z_{Gb} = Tr_b \exp\left(-\beta\widehat{H_b} + \beta\Phi\widehat{Q_b} + \beta\Omega\widehat{J_b}\right) \tag{15}$$

In LQG, quantum black holes are represented by spin network, collection of graphs with links and vertices. For black holes with large area, the major contribution to the entropy comes from the lowermost spins. Hence, only spin 1/2 contribution for *all punctures* is taken into account which yields $A \sim N$ for a total of $N, N \gg 1$ punctures on the horizon. This leads to the equispaced area spectrum as an *approximation*. Of course the higher spins contribute, but their contribution is exponentially suppressed.

So, spectrum of the boundary Hamiltonian is a function of the discrete area spectrum. But the complete spectrum of the boundary Hamiltonian operator is still unknown in LQG. So, we will *assume* that the spectrum of the boundary Hamiltonian operator is also a function of the discrete charge spectrum and the discrete angular momentum spectrum associated with the horizon, respectively.[2] Quantum mechanically, the total charge of a black hole has to be proportional to some fundamental charge, i.e. the black hole is made of such charge particle. Hence the charge spectrum is taken to be equispaced due to quantization [17–19, 35, 36]. In fact angular momentum spectrum can also be considered as equispaced in the macroscopic spectrum limit of the black hole [37], in which we are ultimately interested.

It has already been shown in Subsection (II) that area, charge and angular momentum operators of a black hole commute among them. This implies that they

---

[2] Actually this second assumption follows from the discussion in Subsection (III.A) [16, 32] for spacetimes admitting weakly isolated horizons where there exists a mass function determined by the area and charge associated with the horizon. This is an extension of that assumption to the quantum domain.

are simultaneously diagonalizable. Therefore working in a basis in which area, charge and angular momentum operators are simultaneously diagonal, the partition function (15) can be written as

$$Z_G = \sum_{k,l,m} g(k,l,m) \quad \exp(-\beta(E(A_k, Q_l, J_m) - \Phi Q_l - \Omega J_m)) \tag{16}$$

where $g(k,l,m)$ is the degeneracy corresponding to the area eigenvalue A_k, the charge eigenvalue Q_l and the angular momentum eigenvalue J_m. k,l,m are the quantum numbers corresponding to eigenvalues of area, charge and angular momentum, respectively. In the macroscopic spectra limit of quantum isolated horizons, i.e. regime of the large area, charge and angular momentum eigenvalues $(k \gg 1, l \gg 1, m \gg 1)$, application of the Poisson resummation formula (9) gives

$$Z_G = \int dx\,dy\,dz\,g(A(x), Q(y), J(z)) \exp(-\beta(E(A(x), Q(y), J(z)) - \Phi Q(y) - \Omega J(z))) \tag{17}$$

where x, y, z are, respectively, the continuum limit of k, l, m, respectively.

Now, A, Q and J are, respectively, functions of x, y and z alone. Therefore we have

$$dx = \frac{dA}{A_x}, \quad dy = \frac{dQ}{Q_y}, \quad dz = \frac{dJ}{J_z}$$

where $A_x \equiv \frac{dA}{dx}$ and so on.

So, the partition function, in terms of area, charge and angular momentum as free variables, can be written as follows:

$$Z_G = \int dA \quad dQ \quad dJ \quad \exp[S(A) - \beta(E(A,Q,J) - \Phi Q - \Omega J)], \tag{18}$$

where, following [38], the *microcanonical* entropy of the horizon is defined by $\exp S(A) \equiv \frac{g(A(x), Q(y), J(z))}{\frac{dA dQ dJ}{dx\,dy\,dz}}$ and is a function of horizon area (A) alone, as has been established within LQG [5, 6, 15] .

4. Stability against Gaussian fluctuations

4.1 Saddle-point approximation

The equilibrium configuration of a black hole is given by the saddle point $(\overline{A}, \overline{Q}, \overline{J})$ in the three-dimensional space of integration over area, charge and angular momentum. The idea now is to examine the grand canonical partition function for fluctuations $a = (A - \overline{A}), q = (Q - \overline{Q}), j = (J - \overline{J})$ around the saddle point, in order to determine the stability of the equilibrium isolated horizon under Hawking radiation. We restrict our attention to Gaussian fluctuations, as per common practice in equilibrium statistical mechanics, with the motivation towards extremizing the free energy for the most probable configuration. Taylor expanding Eq. (18) about the saddle point yields

$$Z_G = \exp\left[S(\overline{A}) - \beta M(\overline{A},\overline{Q},\overline{J}) + \beta\Phi\overline{Q} + \beta\Omega\overline{J}\right]$$
$$\times \int da\, dq\, dj\, \exp\{-\frac{\beta}{2}[\left(M_{AA} - \frac{S_{AA}}{\beta}\right)a^2 + (M_{QQ})q^2 + (2M_{AQ})aq \quad (19)$$
$$+(M_{JJ})j^2 + (2M_{AJ})aj + (2M_{QJ})qj]\},$$

where $M(\overline{A},\overline{Q},\overline{J})$ is the mass of equilibrium isolated horizon. Here $M_{AQ} \equiv \left.\frac{\partial^2 M}{\partial A \partial Q}\right|_{(\overline{A},\overline{Q},\overline{J})}$, etc.

We assume, just like in LQG, observables used here are self-adjoint operators over the boundary Hilbert space, and hence their eigenvalues are real [3]. It suffices therefore to restrict integrations over the spectra of these operators to the real axes.

Now, in the saddle-point approximation, the coefficients of terms linear in a, q, j vanish by definition of the saddle point. These imply that

$$\beta = \frac{S_A}{M_A}, \Phi = M_Q, \Omega = M_J \quad (20)$$

Of course these derivatives are calculated at the saddle point.

4.2 Quantum correction of black hole entropy

Note that in the stability criteria derived in the last section, first- and second-order derivatives of the microcanonical entropy of the horizon at equilibrium play a crucial role, in making some of the criteria nontrivial. Thus, corrections to the microcanonical entropy beyond the Bekenstein-Hawking area law, arising due to quantum spacetime fluctuations, might play a role of some significance. It has been shown that [15] the microcanonical entropy for *macroscopic* isolated horizons has the form:

$$S = S_{BH} - \frac{3}{2}\log S_{BH} + O\left(S_{BH}^{-1}\right) \quad (21)$$

$$S_{BH} = \frac{A}{4A_P}, A_P \equiv \text{Planck area}, A \equiv \text{black hole area} \quad (22)$$

In Ref. [15] the above formula was derived for nonrotating, uncharged black holes in (3 + 1) spacetime dimension. But it has already been shown that the above formula equally holds in the case of black holes with charge [33]. Actually black hole entropy depends on the degrees of freedom on its horizon. It is purely a geometrical property of the isolated horizon. Adding charge to the black hole does not alter this geometry at all. In fact it is also shown that results from analysis for isolated horizons with charge is similar to that with angular momentum, except for certain technical issues [33, 35, 39]. Therefore the above formula will be taken to be valid for charged, rotating black holes as well.

4.3 Stability criteria

Convergence of the integral (19) implies that the Hessian matrix (H) has to be positive definite, where

$$H = \begin{pmatrix} \beta M_{AA}(\overline{A},\overline{Q},\overline{J}) - S_{AA}(\overline{A}) & \beta M_{AQ}(\overline{A},\overline{Q},\overline{J}) & \beta M_{AJ}(\overline{A},\overline{Q},\overline{J}).1in \\ \beta M_{AQ}(\overline{A},\overline{Q},\overline{J}) & \beta M_{QQ}(\overline{A},\overline{Q},\overline{J}) & \beta M_{JQ}(\overline{A},\overline{Q},\overline{J}).1in \\ \beta M_{AJ}(\overline{A},\overline{Q},\overline{J}) & \beta M_{JQ}(\overline{A},\overline{Q},\overline{J}) & \beta M_{JJ}(\overline{A},\overline{Q},\overline{J}) \end{pmatrix} \tag{23}$$

The necessary and sufficient conditions for a real symmetric square matrix to be positive definite are that determinants of all principal square submatrices and the determinant of the full matrix are positive [40–42]. This condition leads to the following 'stability criteria:

$$M_{AA}(\overline{A},\overline{Q},\overline{J}) - \frac{S_{AA}(\overline{A})}{\beta} > 0 \tag{24}$$

$$M_{QQ}(\overline{A},\overline{Q},\overline{J}) > 0 \tag{25}$$

$$M_{JJ}(\overline{A},\overline{Q},\overline{J}) > 0 \tag{26}$$

$$M_{QQ}(\overline{A},\overline{Q},\overline{J})M_{JJ}(\overline{A},\overline{Q},\overline{J}) - \left(M_{JQ}(\overline{A},\overline{Q},\overline{J})\right)^2 > 0 \tag{27}$$

$$M_{JJ}(\overline{A},\overline{Q},\overline{J})\left(M_{AA}(\overline{A},\overline{Q},\overline{J}) - \frac{S_{AA}(\overline{A})}{\beta}\right) - \left(M_{AJ}(\overline{A},\overline{Q},\overline{J})\right)^2 > 0 \tag{28}$$

$$M_{QQ}(\overline{A},\overline{Q},\overline{J})\left(M_{AA}(\overline{A},\overline{Q},\overline{J}) - \frac{S_{AA}(\overline{A})}{\beta}\right) - \left(M_{AQ}(\overline{A},\overline{Q},\overline{J})\right)^2 > 0 \tag{29}$$

$$\begin{aligned} &\left[\left(M_{AA}(\overline{A},\overline{Q},\overline{J}) - \frac{S_{AA}(\overline{A})}{\beta}\right)\left(M_{QQ}(\overline{A},\overline{Q},\overline{J})M_{JJ}(\overline{A},\overline{Q},\overline{J}) - \left(M_{JQ}(\overline{A},\overline{Q},\overline{J})\right)^2\right)\right. \\ &-M_{AQ}(\overline{A},\overline{Q},\overline{J})\left(M_{AQ}(\overline{A},\overline{Q},\overline{J})M_{JJ}(\overline{A},\overline{Q},\overline{J}) - M_{JQ}(\overline{A},\overline{Q},\overline{J})M_{AJ}(\overline{A},\overline{Q},\overline{J})\right) \\ &\left.+M_{AJ}(\overline{A},\overline{Q},\overline{J})\left(M_{AQ}(\overline{A},\overline{Q},\overline{J})M_{JQ}(\overline{A},\overline{Q},\overline{J}) - M_{QQ}(\overline{A},\overline{Q},\overline{J})M_{AJ}(\overline{A},\overline{Q},\overline{J})\right)\right] > 0 \end{aligned} \tag{30}$$

Of course, (inverse) temperature β is assumed to be positive for a stable configuration.

Now, the temperature is defined as $T \equiv \frac{1}{\beta} = \frac{M_A}{S_A}$ (from Eq. (20)).

Eqs. (21) and (22) together yield

$$S_A = \frac{1}{4A_P} - \frac{3}{2A} \tag{31}$$

This is positive for macroscopic black holes ($A > > A_P$). So, positivity of M_A implies the positivity of β for macroscopic black holes. The relation $T = \frac{M_A}{S_A}$ implies that

$$\frac{dT}{dA} = \frac{\beta M_A}{(S_A)^2}\left(M_{AA} - \frac{S_{AA}}{\beta}\right) \tag{32}$$

So, the positivity of the quantity $\left(M_{AA} - \frac{S_{AA}}{\beta}\right)$, which is a stability criteria (24), means the positivity of $\frac{dT}{dA}$. In words, a stable black hole becomes hotter as it grows in size. If this is violated, as, for example, in the case of the standard Schwarzschild black hole ([9]), thermal instability is inevitable.

Eq. (20) implies that $M_{QQ} = \frac{d\Phi}{dQ}$. So, positivity of M_{QQ} is an artefact of the fact that accumulation of charge increases the electric potential of the black hole. This is a feature of a stable black hole (25).

Similarly, Eq. (20) shows that $M_{JJ} = \frac{d\Omega}{dJ}$. So, positivity of M_{JJ} implies that the gathering of angular momentum helps the black hole to rotate faster. Hence this is the case with stable black holes (26).

The convexity property of the entropy follows from the condition of convergence of partition function under Gaussian fluctuations [9, 31, 38]. The thermal stability is related to the convexity property of entropy. Hence, the above conditions are correctly the conditions for thermal stability. For chargeless, nonrotating horizons, Eq. (24) reproduces the thermal stability criterion and condition of positive specific heat (i.e. variation of black hole mass with temperature) given in ([13]), as expected. Actually for a chargeless, nonrotating black hole, both the mass and the temperature are functions of the horizon area (A) only. So, the specific heat (C) of the black hole is given as

$$C \equiv \frac{dM}{dT} = \frac{(S_A)^2}{(\beta M_{AA} - S_{AA})} \tag{33}$$

For charged, nonrotating black holes, Eqs. (24), (25) and (29) describe the stability, in perfect agreement with [12], while (24), (26) and (28) describe the thermal stability criteria for uncharged rotating radiant horizons. The new feature for black holes with both charge and angular momentum is that not only does the specific heat have to be positive for stability, but the charge and the angular momentum play important roles as well.

5. Thermal stability of higher-dimensional black holes with arbitrary hairs

5.1 Thermal holography

In this section, we present a generalization of thermal holography for rotating electrically charged quantum radiant horizons discussed in [43], to the situation when the horizon has arbitrary number of hairs [44]. This section of the chapter will of course have substantial overlap with some of the appropriate previous sections of this chapter, so for brevity we focus on the novel aspects here.

5.1.1 Mass associated with horizon

Isolated horizons (b) represent black holes at equilibrium. These isolated horizons are the internal boundaries of spacetime. The first law associated with isolated horizon (b) comes from the Hamiltonian evolution of this spacetime. The law is given as

$$\delta E_h^t = \frac{\kappa^t}{8\pi} \delta A_h + P_i^t \delta C_h^i \tag{34}$$

Here, Einstein summation convention is used, i.e. summation over repeated indices i from 1 to n (=total number of hairs) is implied. E_h^t is the energy function associated with the horizon. κ^t and P_i^t are, respectively, the surface gravity associated with the area of horizon (A_h) and the potential corresponding to the charge (hair) C_h^i. For example, if C_h^i is the angular momentum (J_h), then P_i^t will be the angular velocity (Ω^t). The label 't' denotes the particular time evolution field (t^μ) associated with the spatial hypersurface chosen. E_h^t is assumed here to be a function of A_h and all C_h^i.

The advantage of the isolated (and also the radiant or *dynamical*) horizon description is that one can associate with it a mass M_h^t, related to the ADM energy of the spacetime through the relation

$$E_{ADM}^t = M_h^t + E_{rad}^t \tag{35}$$

where E_{rad}^t is the energy associated with spacetime between the horizon and asymptopia. An isolated horizon admits $E_{rad}^t \neq 0$, and hence a mass is defined *locally* on the horizon.

5.1.2 Quantum algebra and quantum geometry

We consider a quantum black hole with n charges (hairs) $C^1, \ldots, C^n$. These charges are independent of each other. Therefore, respectively, the corresponding operators $\widehat{C^1}, \ldots, \widehat{C^n}$ are also independent of each other, i.e.

$$\left[\widehat{C^i}, \widehat{C^j}\right] = 0, \quad for \quad i \neq j \tag{36}$$

These charges are intrinsic to the black holes and independent of the horizon area (A) of the black hole, if we choose the mass (M) of the black hole to be a dependent variable which depends on the horizon area and the charges. This implies

$$\left[\widehat{A}, \widehat{C^i}\right] = 0, \quad \forall i = 1(1)n \tag{37}$$

where $\widehat{A}$ is the area operator of the black hole.

Choosing mass (M) to be the dependent variable implies that mass (M) is a function of area (A) and all the charges ($C^1, \ldots, C^n$), i.e. $M = M(A, C^1, \ldots, C^n)$. This gives

$$\left[\widehat{M}, \widehat{C^i}\right] = 0, \quad \left[\widehat{M}, \widehat{A}\right] = 0 \quad \forall i = 1(1)n \tag{38}$$

where $\widehat{M}$ is the mass operator of the black hole. Eqs. (36)–(38) together imply that a black hole with a given mass can simultaneously be an eigenstate of its area operator ($\widehat{A}$) and all the charge operators ($\widehat{C^1}, \ldots, \widehat{C^n}$). So, we can consider that a black hole of given mass M has specified area A and specified charges $C^1, \ldots, C^n$.

The Hilbert space of a generic quantum spacetime is given as $\mathcal{H} = \mathcal{H}_b \otimes \mathcal{H}_v$, where $b(v)$ denotes the boundary (bulk) space. A generic quantum state is thus given as

$$|\Psi\rangle = \sum_{b,v} C_{b,v} |\chi_b\rangle \otimes |\psi_v\rangle \tag{39}$$

Now, the full Hamiltonian operator ($\widehat{H}$), operating on $\mathcal{H}$, is given by

$$\widehat{H}|\Psi\rangle = \left(\widehat{H_b} \otimes I_v + I_b \otimes \widehat{H_v}\right)|\Psi\rangle \tag{40}$$

where, respectively, $I_b(I_v)$ is the identity operator on $\mathcal{H}_b(\mathcal{H}_v)$ and $\widehat{H_b}\left(\widehat{H_v}\right)$ is the Hamiltonian operator on $\mathcal{H}_b(\mathcal{H}_v)$.

Now, the bulk Hamiltonian operator annihilates bulk physical states:

$$\widehat{H_v}|\psi_v\rangle = 0 \tag{41}$$

The charge operators $\widehat{C^i}$s are each an infinitesimal generator of a continuous transformation on the bulk Hilbert space. For example, electric charge operator ($\widehat{Q}$) is the generator of local $U(1)$ transformation, and angular momentum operator ($\widehat{J}$) is the generator of local spatial rotation. So, the assumption that bulk quantum spacetime is invariant under all these transformations implies that bulk spacetime is free of any charge (hair). This gives

$$\widehat{C^i_v}|\psi_v\rangle = 0 \tag{42}$$

where $\widehat{C^i_v}$ is the bulk charge operator corresponding to the charge C^i.

So Eqs. (41) and (42) together produce

$$\left[\widehat{H_v} - \beta P_i \widehat{C^i_v}\right]|\psi_v\rangle = 0. \tag{43}$$

where β can be any function, but we treat it as inverse temperature of the black hole afterwards.

5.2 Grand canonical partition function

We now consider the black hole with the contact of a heat bath, at some (inverse) temperature β, with which it can exchange energy, charge, angular momentum and all quantum hairs. The grand canonical partition function of the black hole is given as

$$Z_G = Tr\left(exp\left(-\beta\widehat{H} + \beta P_i\widehat{C^i}\right)\right) \tag{44}$$

where the trace is taken over all states. This definition, together with Eqs. (39) and (43) yield

$$\begin{aligned} Z_G &= \sum_{b,v} |C_{b,v}|^2 \langle\psi_v|\psi_v\rangle \left\langle\chi_b\right| exp\left(-\beta\widehat{H_b} + \beta P_i\widehat{C^i}\right)\left|\chi_b\right\rangle \\ &= \sum_b |C_b|^2 \left\langle\chi_b\right| exp\left(-\beta\widehat{H_b} + \beta P_i\widehat{C^i}\right)\left|\chi_b\right\rangle, \end{aligned} \tag{45}$$

assuming that the boundary states are normalized. The partition function thus turns out to be completely determined by the boundary states (Z_{Gb}), i.e.

$$Z = Z_{Gb} = Tr_b \exp\left(-\beta\widehat{H} + \beta P_i\widehat{C^i}\right)$$
$$= \sum_{l,k_1,\dots,k_n} g(l,k_1,\dots,k_n)\ \exp\left(-\beta\left(E\left(A_l, C^1_{k_1},\dots,C^n_{k_n}\right) - \sum_{i=1}^{n} P_i C^i_{k_i}\right)\right). \tag{46}$$

where $g(l,k_1,\dots,k_n)$ is the degeneracy corresponding to energy $E\left(A_l, C^1_{k_1},\dots,C^n_{k_n}\right)$ and l,k_i are the quantum numbers corresponding to area and charge C^i, respectively. Here, the spectrum of the boundary Hamiltonian operator is assumed to be a function of area and all other charges of the boundary, considered here to be the horizon. Following [12], it is further assumed that these 'hairs' have a discrete spectrum. In the macroscopic limit of the black hole, they all have large eigenvalues, i.e. $(l,k_i >> 1)$, so that application of the Poisson resummation formula (9) gives

$$Z_G = \int dx \left(\prod_{i=1}^{n}\int dy_i\right) g(A(x), C^1(y_1),\dots,C^n(y_n)) \exp(-\beta(E(A(x),C^1(y_1),\dots,C^n(y_n))) - \sum_{i=1}^{n} P_i C^i(y_i))). \tag{47}$$

where x, y_i are, respectively, the continuum limits of l,k_i, respectively.

Following [12], we now assume that the macroscopic spectrum of the area and all charges are linear in their arguments, so that a change of variables gives, with constant Jacobian, the result

$$Z_G = \int dA \left(\prod_{i=1}^{n}\int dC^i\right) \exp(S(A) - \beta(E(A,C^1,\dots,C^n) - P_iC^i)), \tag{48}$$

where, following [38], the *microcanonical* entropy of the horizon is defined by

$$\exp S(A) \equiv \frac{g(A(x), C(y_1),\dots,C(y_n))}{\frac{dA}{dx}\frac{dC^1}{dy_1}\cdots\frac{dC^n}{dy_n}} \tag{49}$$

5.3 Saddle-point approximation

The equilibrium configuration of black hole is given by the saddle point $(\overline{A},\overline{C}^1,,\overline{C}^n)$ in the $(n+1)$ dimensional space of integration over area and n charges. This configuration is identified with an isolated horizon, as already mentioned. The idea now is to examine the grand canonical partition function for fluctuations $a = (A - \overline{A})$ and $c^i = \left(C^i - \overline{C}^i\right)$ around the saddle point, in order to determine the stability of the equilibrium isolated horizon under Hawking radiation. We restrict our attention to Gaussian fluctuations. Taylor expanding Eq. (48) about the saddle point yields

$$Z_G = \exp\left[S(\overline{A}) - \beta M\left(\overline{A}, \overline{C}^1, \dots, \overline{C}^n\right) + \beta P_i \overline{C}^i\right]$$
$$\times \int dA \left(\prod_{i=1}^{n} \int dC^i\right) \exp\{-\frac{1}{2}[(\beta M_{AA} - S_{AA})a^2 + 2\sum_{i=1}^{n} \beta M_{AC^i} a c^i \tag{50}$$
$$+ \sum_{i=1}^{n}\sum_{j=1}^{n} \beta M_{C^i C^j} c^i c^j]\},$$

where $M\left(\overline{A}, \overline{C}^1, \dots, \overline{C}^n\right)$ is the mass of equilibrium isolated horizon.

Here $M_{AC^i} \equiv \partial^2 M / \partial A \partial C^i \big|_{\left(\overline{A}, \overline{C}^1, \dots, \overline{C}^n\right)}$, etc.

Now, in the saddle-point approximation, the coefficients of terms linear in a, c^i vanish by definition of the saddle point. These imply that

$$\beta = \frac{S_A}{M_A}, \quad P_i = M_{C^i} \tag{51}$$

Of course these are evaluated at the saddle point.

5.4 Stability criteria

Convergence of the integral (50) implies that the Hessian matrix (H) has to be positive definite, where

$$H = \begin{pmatrix} \beta M_{AA} - S_{AA} & \beta M_{AC^1} & \beta M_{AC^2} & \dots\dots\dots & \beta M_{AC^n} \\ \beta M_{AC^1} & \beta M_{C^1C^1} & \beta M_{C^1C^2} & \dots\dots\dots & \beta M_{C^1C^n} \\ \beta M_{AC^2} & \beta M_{C^2C^1} & \beta M_{C^2C^2} & \dots\dots\dots & \beta M_{C^2C^n} \\ \dots\dots & \dots\dots & \dots\dots & \dots\dots\dots & \dots\dots \\ \beta M_{AC^n} & \beta M_{C^nC^1} & \beta M_{C^nC^2} & \dots\dots\dots & \beta M_{C^nC^n} \end{pmatrix} \tag{52}$$

Here, all the derivatives are calculated at the saddle point. Hence the stability criteria, i.e. the criteria for positive definiteness of Hessian matrix, are given as

$$D_1 > 0, D_2 > 0, \dots, D_{n+1} > 0 \tag{53}$$

where

$$D_1 = \beta M_{AA} - S_{AA}, \quad D_2 = \begin{vmatrix} \beta M_{AA} - S_{AA} & \beta M_{AC^1} \\ \beta M_{AC^1} & \beta M_{C^1C^1} \end{vmatrix},$$
$$D_3 = \begin{vmatrix} \beta M_{AA} - S_{AA} & \beta M_{AC^1} & \beta M_{AC^2} \\ \beta M_{AC^1} & \beta M_{C^1C^1} & \beta M_{C^1C^2} \\ \beta M_{AC^2} & \beta M_{C^2C^1} & \beta M_{C^2C^2} \end{vmatrix}, \dots, D_{n+1} = |H| \tag{54}$$

where $|H|$ = determinant of the Hessian matrix H.

Of course, (inverse) temperature β is assumed to be positive for a stable configuration. We again find that temperature must increase with horizon area, inherent in the positivity of the quantity $(\beta M_{AA} - S_{AA})$.

The convexity property of the entropy follows from the condition of convergence of partition function under Gaussian fluctuations [9, 31, 38]. The thermal stability is related to the convexity property of entropy. Hence, the above conditions are correctly the conditions for thermal stability. For rotating charged horizons, Eqs. (53) and (54) reproduce the thermal stability criterion with $n = 2$, i.e. $D_1 > 0, D_2 > 0, D_3 > 0$ with the identification that charge of the black hole (Q)=C^1 and angular momentum of the black hole (J)=C^2. It can be easily checked that these three conditions correctly reproduce the earlier ([43]) seven conditions of thermal stability of charged rotating black holes. Eqs. (53) and (54) necessarily tell us that thermal stability of black hole is a consequence of the interplay among all the charges of the black hole.

Now, we are going to show that Eqs. (53) and (54) correctly produce the criteria of stability for charged rotating black holes (24)–(30), taking $n = 2$.

Consider the following integral,

$$I = \int\int\int dxdydz\exp(-(ax^2 + by^2 + cz^2 + 2dxy + 2eyz + 2fzx))$$

Define $U \equiv (ax^2 + by^2 + cz^2 + 2dxy + 2eyz + 2fzx)$

Now, we can rewrite the argument of the exponential (U) part as

$$U = a\left(x + \frac{d}{a}y + \frac{f}{a}z\right)^2 + \frac{(ab - d^2)}{a}\left(y + \frac{e - df/a}{((ab - d^2)/a)}z\right)^2 + \left(\frac{ac - f^2}{a} - \frac{(e - df/a)^2}{((ab - d^2)/a)}\right)z^2 \tag{55}$$

Considering the notations, given in Eqs. (53) and (54), we can write

$$U = D_1\left(x + \frac{d}{a}y + \frac{f}{a}z\right)^2 + \frac{D_2}{D_1}\left(y + \frac{e - df/a}{((ab - d^2)/a)}z\right)^2 + \frac{D_3}{D_2}z^2 \tag{56}$$

where

$$\begin{aligned} \mathrm{D_1} &= \mathrm{a} \\ \mathrm{D_2} &= (\mathrm{ab} - \mathrm{d}^2) \\ \frac{\mathrm{D_3}}{\mathrm{D_2}} &= \frac{(\mathrm{abc} - \mathrm{cd}^2 - \mathrm{bf}^2 - \mathrm{ae}^2 + 2\mathrm{dfe})}{(\mathrm{ab} - \mathrm{d}^2)} \\ &= \frac{\mathrm{ac} - \mathrm{f}^2}{\mathrm{a}} - \frac{(\mathrm{e} - \mathrm{df/a})^2}{((\mathrm{ab} - \mathrm{d}^2)/\mathrm{a})} \end{aligned} \tag{57}$$

So, we have

$$I = \int\int\int dxdydz\exp\left(-\left(D_1\left(x + \frac{d}{a}y + \frac{f}{a}z\right)^2 + \frac{D_2}{D_1}\left(y + \frac{e - df/a}{((ab - d^2)/a)}z\right)^2 + \frac{D_3}{D_2}z^2\right)\right) \tag{58}$$

Consider the following change of variables:

$$\begin{pmatrix} \underline{x} \\ \underline{y} \\ \underline{z} \end{pmatrix} = \begin{pmatrix} 1 & \frac{d}{a} & \frac{f}{a} \\ 0 & 1 & \frac{e - df/a}{((ab - d^2)/a)} \\ 0 & 0 & 1 \end{pmatrix} \begin{pmatrix} x \\ y \\ z \end{pmatrix} \tag{59}$$

Therefore Eqs. (58) and (59) together give

$$I = \int\int\int d\underline{x} d\underline{y} d\underline{z} \quad A \cdot exp\left(-\left(D_1 \underline{x}^2 + \frac{D_2}{D_1}\underline{y}^2 + \frac{D_3}{D_2}\underline{z}^2\right)\right) \tag{60}$$

where A is the Jacobian of the transformation matrix, i.e.

$$A = \begin{vmatrix} 1 & \frac{d}{a} & \frac{f}{a} \\ 0 & 1 & \frac{e - df/a}{((ab - d^2)/a)} \\ 0 & 0 & 1 \end{vmatrix} = 1 \tag{61}$$

$$I = \int\int\int d\underline{x} d\underline{y} d\underline{z} \quad exp\left(-\left(D_1 \underline{x}^2 + \frac{D_2}{D_1}\underline{y}^2 + \frac{D_3}{D_2}\underline{z}^2\right)\right) \tag{62}$$

This expression explicitly shows that I will be converging if and only if $D_1 > 0, D_2 > 0, D_3 > 0$. This is what we have claimed in this section as the condition for thermal stability of rotating charged black holes.

From the expression (57), we get:

1) If $D_1 > 0, D_2 > 0$, then $b > 0$.

2) If $D_1 > 0, D_2 > 0, D_3 > 0$, then $(ac - f^2) > 0$. Consequently, $c > 0$.

3) The expression of D_3 can be rearranged as

$$\begin{aligned} D_3 &= (abc - cd^2 - bf^2 - ae^2 + 2dfe) \\ &= (ab^2c - bcd^2 - b^2f^2 - abe^2 + 2bdfe)/b \\ &= \left((ab - d^2)(bc - e^2) - (bf - ed)^2\right)/b \end{aligned} \tag{63}$$

So, the positivity of b, $D_2 (= ab - d^2)$ and D_3 implies that $(bc - e^2) > 0$. Therefore these conditions for thermal stability described by $D_1 > 0, D_2 > 0, D_3 > 0$ are same as those described by inequalities (24–30) for rotating charged black holes.

For an n dimensional matrix, the total number of submatrices including the whole matrix (N_s) is given as

$$\begin{aligned} N_s &= n_{C_1} + n_{C_2} + \dots \dots + n_{C_n} \\ &= 2^n - 1 \end{aligned} \tag{64}$$

Now, any generic quadratic expression of n variables can be rearranged by redefining variables as an quadratic expression without any cross term, i.e. of the form $\sum_{i=1}^{n} a_i x_i^2$.

Thus if we consider the positivity of determinants of the all submatrices of Hessian (including itself), then we have to check $(2^{n+1} - 1)$ conditions for testing thermal stability of a black hole with n charges. On the other hand, if we follow the different procedure set in this chapter, then $(n + 1)$ conditions have to be checked for testing thermal stability of a black hole with n charges. Obviously $(2^{n+1} - 1)$ is greater than $(n + 1)$ for $n \geq 1$, i.e. black hole with at least one charge. But there are certain advantages of checking these additional criteria, i.e. positivity of submatrices of Hessian matrix. This is very useful for studying 'Quasi Stable' black holes, especially for studying the fluctuations of charges for such black holes [45]. But this is beyond the scope of this chapter. In fact the issue of thermal fluctuations for stable black holes is also interesting [46]. A stable black hole has to satisfy all the stability criteria. So, a simple inequality, i.e. determinant of a submatrix of Hessian matrix, may be negative. This can be easily checked, and the corresponding black hole is concluded to be unstable under Hawking radiation.

6. Discussions

The novelty of our approach is that it is purely based on quantum aspects of spacetime. *Classical metric* has not been used anywhere in the analysis. The construction of the partition function is based on LQG, e.g. the use of Chern-Simons states, the splitting up of the total Hilbert space, etc. and also on the Hamiltonian formulation of spacetimes admitting weakly isolated horizons. The entropy correction also follows from the quantum theory.

In this analysis of thermal stability of black holes, two physically reasonable assumptions are made. In classical Hamiltonian GR, total Hamiltonian vanishes. So, it is considered that the total quantum Hamiltonian operator annihilates the bulk states of quantum matter coupled spacetime. A similar argument follows for the assumption of the quantum constraint on the volume charge operator. These two assumptions may be considered to be one due to their fundamental similarity, and they ultimately give rise to a single quantum constraint.

In Section (III.C), a second assumption is made regarding the eigenvalue spectrum of the energy of the black hole. The classical mass associated with the horizon is a function of horizon area, charge and angular momentum. These horizon area, charge and angular momentum are the functions of the local fields on the horizon. So, quantization of the classical horizon area, charge and angular momentum will definitely lead to a well-defined boundary Hamiltonian operator. The existence of a quantum boundary Hamiltonian operator, acting on the boundary Hilbert space of the black hole, is an assumption as the exact form of such a Hamiltonian operator is still unknown. But the fact that its eigenvalue spectrum is a function of eigenvalue spectra of the area, charge and angular momentum operators are an obvious assumption, as it is bound to happen if such a boundary Hamiltonian operator exists. It follows from the classical analogue—the mass associated with the horizon must be a function of the horizon area, charge and angular momentum for a consistent Hamiltonian evolution.

In this chapter, we have derived the criteria for thermal stability of charged rotating black holes, for horizon areas that are largely relative to the Planck area (in these dimensions). We also generalize it for black holes with arbitrary hairs in any spacetime dimension. Like earlier, results of LQG and equilibrium statistical mechanics of the grand canonical ensemble are sufficient for our analysis. The only assumption is that the mass of the black hole is a function of its horizon area and all the hairs. The obtained stability criteria can be applied to check the thermal stability of any black hole whose mass is given as function of its charges, and in fact this has been done [43] for various black holes as well.

Author details

Aloke Kumar Sinha
Haldia Government College, West Bengal, India

*Address all correspondence to: akshooghly@gmail.com

References

[1] Hawking SW, Page DN. Thermodynamics of black holes in anti-de sitter space. Communications in Mathematical Physics. 1983;**87**:577

[2] Davis PCW. The thermodynamic theory of black holes. Proceedings of the Royal Society. 1977;**A353**:499

[3] Rovelli C. Quantum Gravity, Cambridge Monographs on Mathematical Physics. UK: Cambridge University Press; 2004

[4] Thiemann T. Modern Canonical Quantum General Relativity, Cambridge Monographs on Mathematical Physics. UK: Cambridge University Press; 2007

[5] Ashtekar A, Baez J, Corichi A, Krasnov K. Quantum geometry and black hole entropy. Physical Review Letters. 1998;**80**:904; e-Print arXiv: gr-qc/9710007

[6] Ashtekar A, Baez J, Krasnov K. Quantum geometry of isolated horizons and black hole entropy. Advances in Theoretical and Mathematical Physics. 2000;**4**:1; e-Print arXiv : gr-qc/ 0005126

[7] Kaul RK, Majumdar P. Quantum black hole entropy. Physics Letters. 1998;**B439**:267; e-Print arXiv : gr-qc/ 9801080

[8] Basu R, Kaul RK, Majumdar P. Entropy of isolated horizons revisited. Physics Review. 2010;**D82**:024007; e-Print arXiv : 0907.0846v3

[9] Chatterjee A, Majumdar P. Universal canonical black hole entropy. Physical Review Letters. 2004;**92**:141031; e-Print arXiv : gr-qc/0309026

[10] Chatterjee A, Majumdar P. Mass and charge fluctuations and black hole entropy. Physics Review. 2005;**D71**: 024003; e-Print arXiv : gr-qc/0409097

[11] Chatterjee A, Majumdar P. Universal criterion for black hole stability. Physics Review. 2005;**D72**: 044005; e-Print arXiv : gr-qc/0504064

[12] Majhi A, Majumdar P. Charged quantum black holes : Thermal stability criterion. Classical and Quantum Gravity. 2012;**29**:135013; e-Print arXiv : 1108.4670

[13] Majumdar P. Generalized hawking-page phase transition. Classical and Quantum Gravity. 2007;**24**:1747; e-Print arXiv : gr-qc/ 0701014

[14] Majumdar P. Holography, Gauge-gravity connection and black hole entropy. International Journal of Modern Physics A. 2009;**A24**:3414; e-Print arXiv :0903.5080

[15] Kaul RK, Majumdar P. Logarithmic correction to the bekenstein-hawking entropy. Physical Review Letters. 2000;**84**:5255; e-Print arXiv : gr-qc/ 0002040

[16] Ashtekar A, Krishnan B. Isolated and dynamical horizons and their applications. Living Reviews in Relativity. 2004;**7**:10; e-Print arXiv : gr-qc/ 0407042

[17] Kaul RK, Rama SK. Black hole entropy from spin one punctures. Physical Review D. 2003;**68**:024001, e-Print arxiv: gr-qc/0301128v1

[18] Frodden E, Perez A, Pranzetti D, Roken C. Modelling black holes with angular momentum in loop quantum gravity. General Relativity and Gravitation. 2014;**46**:1828, e-Print arxiv: 1212.5166

[19] Achour JB, Noui K, Perez A. Analytic continuation of the rotating black hole state counting. JHEP 1608. 2016:149, e-Print arxiv: 1607.02380

[20] Emparan R, Reall HS. Black holes in higher dimensions. Living Reviews in Relativity. 2008;**11**:6

[21] Carter B. Axisymmetric black hole has only two degrees of freedom. Physical Review Letters. 1971;**26**(6): 331-333

[22] Coleman S, Preslill J, Wilczek F. Growing hair on black holes. Physical Review Letters. 1975;**67**:1978

[23] Bizon P. Colored black holes. Physical Review Letters. 1990;**64**:2844

[24] Krauss LM, Wilczek F. Discrete gauge symmetry in continuum theories. Physical Review Letters. 1989;**62**:1221

[25] Bowick MJ, Giddings SB, Harvey JA, Horowitz GT, Strominger A. Axionic black holes and an aharonov-bohm effect for strings. Physical Review Letters. 1988;**61**:2823

[26] Lee K, Nair VP, Weinberg EJ. The fate of magnetically charged black holes. Physical Review D. 1992;**45**:2751

[27] Preskill J. Quantum hair. Physica Scripta. 1991;**T36**:258-264

[28] Coleman S, Preslill J, Wilczek F. Dynamical effect of quantum hair. Modern Physics Letters A. 1991;**06**:1631

[29] Myers RC, Perry MJ. Black holes in higher dimensional space-times. Annals of Physics. 1986;**172**:304-347

[30] Das S, Majumdar P, Bhaduri RK. General logarithmic corrections to black hole entropy. Classical and Quantum Gravity. 2002;**19**:2355-2368, e-Print arxiv:hep-th/0111001

[31] R. Monteiro. Classical and thermodynamic stability of black holes. 2010. e-Print arxiv:hep-th 1006.5358

[32] Ashtekar A, Fairhurst S, Krishnan B. Isolated horizons: Hamiltonian evolution and the first law. Physics Review, D62. 2000:104025; e-Print arxiv:gr-qc/0005083v3

[33] Ashtekar A, Beetle C, Fairhurst S. Mechanics of isolated horizons. Classical and Quantum Gravity. 2000;**17**:253-298, e-Print arXiv:gr-qc/9907068

[34] Ashtekar A, Pawlowski T, Broeck C. Mechanics of higher-dimensional black holes in asymptotically anti-de sitter space-times. Classical and Quantum Gravity. 2007;**24**:625-644; e-Print arXiv:gr-qc/0611049v2

[35] Ashtekar A, Beetle C, Lewandowski J. Mechanics of rotating isolated horizons. PRD 64. 2001: 044016, e-print arXiv:gr-qc/0103026

[36] Majhi A, Majumdar P. Quantum hairs and entropy of quantum isolated horizon from chern-simons theory. Classical and Quantum Gravity. 2014; **31**:195003; e-Print arXiv: 1301.4553

[37] Gour G, M AJ. Medved. Kerr black hole as a quantum rotator. Classical and Quantum Gravity. 2003:**20**;2261-2274. e-Print arXiv:gr-qc/0211089v1

[38] Landau LD, Lifschitz EM. Statistical Physics. Oxford, England: Pergamon Press; 1980

[39] Ashtekar A, Engle J, Broeck CVD. Quantum horizons and black hole entropy: Inclusion of distortion and rotation. Classical and Quantum Gravity. 2005;**22**:L27-L34. e-Print arxiv: gr-qc/0412003v1

[40] Meyer CD. Matrix Analysis and Applied Linear Algebra. Society for Industrial and Applied Mathematics; Har/Cdr edition. Philadelphia: University City Science Center; 1 June 2000

[41] Bhatia R. Positive Definite Matrices. USA: Princeton University Press; 2015. ISBN: 9780691168258

[42] Pinkus A. Totally Positive Matrices. Cambridge University Press; 2009

[43] Sinha AK, Majumdar P. Thermal stability of charged rotating quantum black holes. Modern Physics Letters A. 2017;**32**(37):1750208; e-Print Arxiv: 1512.04181v2

[44] Sinha AK. Thermal stability of black holes with arbitrary hairs. Modern Physics Letters A. 2018;**33**:1850031; e-Print arxiv:1702.08175

[45] Sinha AK. Quasi stable black holes and their implications. Classical and Quantum Gravity. 2018;**36**(2019): 035003

[46] Sinha AK. Thermal fluctuations and correlations among hairs of a stable quantum black hole: Some examples. Modern Physics Letters A. 2018:33, 1850190

Chapter 4

The Black Hole Binary Gravitons and Related Problems

Miroslav Pardy

Abstract

The energy spectrum of graviton emitted by the black hole binary is calculated in the first part of the chapter. Then, the total quantum loss of energy is calculated in the Schwinger theory of gravity. In the next part, we determine the electromagnetic shift of energy levels of H-atom electrons by calculating an electron coupling to the black hole thermal bath. The energy shift of electrons in H-atom is determined in the framework of nonrelativistic quantum mechanics. In the last section, we determine the velocity of sound in the black hole atmosphere, which is here considered as the black hole photon sea. Derivation is based on the thermodynamic theory of the black hole photon gas.

Keywords: graviton, Schwinger source theory, spectrum of H-atom, Coulomb potential, black hole spectrum, energy shift, sound

1. The graviton spectrum of the black hole binary

In 1916, Schwarzschild published the solution of the Einstein field equations [1] that were later understood to describe a black hole [2, 3], and in 1963, Kerr generalized the solution to rotating black holes [4]. The year 1970 was the starting point of the theoretical work leading to the understanding of black hole quasinormal modes [5–7], and in the 1990s, higher-order post-Newtonian calculations [8] were performed and later the extensive analytical studies of relativistic two-body dynamics were realized [9, 10]. These advances, together with numerical relativity breaks through in the past decade [11–13]. Numerous black hole candidates have now been identified through electromagnetic observations [14–16]. The black hole binary and their rotation and mergers are open problem of the astrophysics, and it is the integral part of the binary black hole physics.

The binary pulsar system PSR B**1913+16** (also known as PSR J1915+1606) discovered by Hulse and Taylor [17] and subsequent observations of its energy loss by Taylor and Weisberg [18] demonstrated the existence of gravitational waves [19].

By the early 2000s, a set of initial detectors was completed, including TAMA 300 in Japan, GEO600 in Germany, the Laser Interferometer Gravitational-Wave Observatory (LIGO) in the United States, and Virgo in Italy. In 2015, Advanced LIGO became the first of a significantly more sensitive network of advanced detectors (a second-generation interferometric gravitational wave detector) to begin observations [20].

Taylor and Hulse, working at the Arecibo Radiotelescope, discovered the radio pulsar PSR B**1913+16** in a binary, in 1974, and this is now considered as the best general relativistic laboratory [21].

Pulsar PSR B**1913+16** is the massive body of the binary system where each of the rotating pairs is 1.4 times the mass of the Sun. These neutron stars rotate around each other in an orbit not much larger than the Sun's diameter, with a period of 7.8 h. Every 59 ms, the pulsar emits a short signal that is so clear that the arrival time of a 5 min string of a set of such signals can be resolved within 15 μs.

A pulsar model based on strongly magnetized, rapidly spinning neutron stars was soon established as consistent with most of the known facts [22]; its electrodynamical properties were studied theoretically [23] and shown to be plausibly capable of generating broadband radio noise detectable over interstellar distances. The binary pulsar PSR B**1913+16** is now recognized as the harbinger of a new class of unusually short-period pulsars, with numerous important applications.

Because the velocities and gravitational energies in a high-mass binary pulsar system can be significantly relativistic, strong-field and radiative effects come into play. The binary pulsar PSR B**1913+16** provides significant tests of gravitation beyond the weak-field, slow-motion limit [24, 25].

We do not repeat here the derivation of the Einstein quadrupole formula in the Schwinger gravity theory [26]. We show that just in the framework of the Schwinger gravity theory, it is easy to determine the spectral formula for emitted gravitons and the quantum energy loss formula of the binary system. The energy loss formula is general, including black hole binary, and it involves arbitrarily strong gravity.

Since the measurement of the motion of the black hole binaries goes on, we hope that sooner or later the confirmation of our formula will be established.

1.1 The Schwinger approach for the problem

Source methods by Schwinger are adequate for the solution of the calculation of the spectral formula of gravitons and energy loss of binary. Source theory [27, 28] was initially constructed to describe the particle physics situations occurring in high-energy physics experiments. However, it was found that the original formulation simplifies the calculations in the electrodynamics and gravity, where the interactions are mediated by photon and graviton, respectively. The source theory of gravity forms the analogue of quantum electrodynamics because, while in QED the interaction is mediated by the photon, the gravitational interaction is mediated by the graviton [29]. The basic formula in the source theory is the vacuum-to-vacuum amplitude [30]:

$$\langle 0_+|0_-\rangle = e^{\frac{i}{\hbar}W(S)}, \tag{1}$$

where the minus and plus symbols refer to any time before and after the region of space–time with action of sources. The exponential form is postulated to express the physically independent experimental arrangements, with result that the associated probability amplitudes multiply and the corresponding W expressions add [27, 28].

In the flat space-time, the field of gravitons is described by the amplitude (1) with the action ($c = 1$ in the following text) [31]

$$W(T) = 4\pi G\int (dx)(dx')\left[T^{\mu\nu}(x)D_+(x-x')T_{\mu\nu}(x') - \frac{1}{2}T(x)D_+(x-x')T(x')\right], \tag{2}$$

where the dimensionality of $W(T)$ has the dimension of the Planck constant $\hbar$ and $T_{\mu\nu}$ is the momentum and energy tensor that, for a particle trajectory $\mathbf{x} = \mathbf{x}(t)$, is defined by the equation [32]

$$T^{\mu\nu}(x) = \frac{p^\mu p^\nu}{E}\delta(\ \mathbf{x} - \mathbf{x}(t)), \tag{3}$$

where p^μ is the relativistic four-momentum of a particle with a rest mass m and

$$p^\mu = (E,\ \mathbf{p}) \tag{4}$$

$$p^\mu p_\mu = -m^2, \tag{5}$$

and the relativistic energy is defined by the known relation

$$E = \frac{m}{\sqrt{1-\mathbf{v}^2}}, \tag{6}$$

where $\mathbf{v}$ is the three-velocity of the moving particle.

Symbol $T(x)$ in Eq. (2) is defined as $T = g_{\mu\nu}T^{\mu\nu}$, and $D_+(x - x')$ is the graviton propagator whose explicit form will be determined later.

1.2 The power spectral formula in general

It may be easy to show that the probability of the persistence of vacuum is given by the following formula [27]:

$$|\langle 0_+|0_-\rangle|^2 = \exp\left\{-\frac{2}{\hbar}\ \mathrm{Im\,W}\right\} \stackrel{d}{=} \exp\left\{-\int dtd\omega\frac{1}{\hbar\omega}P(\omega,t)\right\}, \tag{7}$$

where the so-called power spectral function $P(\omega,t)$ has been introduced [27]. For the extraction of the spectral function from Im W, it is necessary to know the explicit form of the graviton propagator $D_+(x - x')$. This propagator involves the graviton property of spreading with velocity c. It means that its mathematical form is identical with the photon propagator form. With regard to Schwinger et al. [33], the x-representation of $D(k)$ in Eq. (2) is as follows:

$$D_+(x-x') = \int\frac{(dk)}{(2\pi)^4}e^{ik(x-x')}D(k), \tag{8}$$

where

$$D(k) = \frac{1}{|\mathbf{k}^2| - (k^0)^2 - i\epsilon}, \tag{9}$$

which gives

$$D_+(x-x') = \frac{i}{4\pi^2}\int_0^\infty d\omega\frac{\sin\omega|\mathbf{x}-\mathbf{x}'|}{|\mathbf{x}-\mathbf{x}'|}e^{-i\omega|t-t'|}. \tag{10}$$

Now, using Eqs. (2), (7), and (10), we get the power spectral formula in the following form:

$$P(\omega,t) = 4\pi G\omega \int (d\ \mathbf{x})(d\ \mathbf{x}')dt' \frac{\sin\omega|\ \mathbf{x}-\mathbf{x}'|}{|\ \mathbf{x}-\ \mathbf{x}'|}\cos\omega(t-t')$$
$$\times\left[T^{\mu\nu}(\ \mathbf{x},t)T_{\mu\nu}(\ \mathbf{x}',t') - \frac{1}{2}g_{\mu\nu}T^{\mu\nu}(\ \mathbf{x},t)g_{\alpha\beta}T^{\alpha\beta}(\ \mathbf{x}',t')\right]. \quad (11)$$

1.3 The power spectral formula for the binary system

In the case of the binary system with masses m_1 and m_2, we suppose that they move in a uniform circular motion around their centre of gravity in the xy plane, with corresponding kinematical coordinates:

$$\mathbf{x}_1(t) = r_1\left(\mathbf{i}\cos(\omega_0 t) + \mathbf{j}\sin(\omega_0 t)\right) \quad (12)$$

$$\mathbf{x}_2(t) = r_2\left(\mathbf{i}\cos(\omega_0 t+\pi) + \mathbf{j}\ \sin(\omega_0 t+\pi)\right) \quad (13)$$

with

$$\mathbf{v}_i(t) = d\mathbf{x}_i/dt, \quad \omega_0 = v_i/r_i, \quad v_i = |\mathbf{v}_i| \quad (i=1,2). \quad (14)$$

For the tensor of energy and momentum of the binary, we have

$$T^{\mu\nu}(x) = \frac{p_1^\mu p_1^\nu}{E_1}\delta(\mathbf{x}-\mathbf{x}_1(t)) + \frac{p_2^\mu p_2^\nu}{E_2}\delta(\mathbf{x}-\mathbf{x}_2(t)), \quad (15)$$

where we have omitted the tensor $t_{\mu\nu}^G$, which is associated with the massless, gravitational field distributed all over space and proportional to the gravitational constant G [32].

After the insertion of Eq. (15) into Eq. (11), we get [33]

$$P_{total}(\omega,t) = P_1(\omega,t) + P_{12}(\omega,t) + P_2(\omega,t), \quad (16)$$

where $(t'-t=\tau)$

$$P_1(\omega,t) = \frac{G\omega}{r_1\pi}\int_{-\infty}^{\infty} d\tau \frac{\sin[2\omega r_1\sin(\omega_0\tau/2)]}{\sin(\omega_0\tau/2)}\cos\omega\tau$$
$$\times\left(E_1^2(\omega_0^2 r_1^2\cos\omega_0\tau - 1)^2 - \frac{m_1^4}{2E_1^2}\right), \quad (17)$$

$$P_2(\omega,t) = \frac{G\omega}{r_2\pi}\int_{-\infty}^{\infty} d\tau \frac{\sin[2\omega r_2\sin(\omega_0\tau/2)]}{\sin(\omega_0\tau/2)}\cos\omega\tau$$
$$\times\left(E_2^2(\omega_0^2 r_2^2\cos\omega_0\tau - 1)^2 - \frac{m_2^4}{2E_2^2}\right), \quad (18)$$

$$P_{12}(\omega,t) = \frac{4G\omega}{\pi}\int_{-\infty}^{\infty} d\tau \frac{\sin\omega\left[r_1^2+r_2^2+2r_1r_2\cos(\omega_0\tau)\right]^{1/2}}{\left[r_1^2+r_2^2+2r_1r_2\cos(\omega_0\tau)\right]^{1/2}}\cos\omega\tau$$
$$\times\left(E_1E_2(\omega_0^2 r_1 r_2\cos\omega_0\tau + 1)^2 - \frac{m_1^2 m_2^2}{2E_1E_2}\right). \quad (19)$$

1.4 The quantum energy loss of the binary

Using the following relations

$$\omega_0\tau = \varphi + 2\pi l, \quad \varphi\in(-\pi,\pi), \quad l = 0,\ \pm1,\ \pm2, \ldots \quad (20)$$

$$\sum_{l=-\infty}^{l=\infty} \cos 2\pi l \frac{\omega}{\omega_0} = \sum_{l=-\infty}^{\infty} \omega_0 \delta(\omega - \omega_0 l), \tag{21}$$

we get for $P_i(\omega, t)$, with ω being restricted to positive:

$$P_i(\omega, t) = \sum_{l=1}^{\infty} \delta(\omega - \omega_0 l) P_{il}(\omega, t). \tag{22}$$

Using the definition of the Bessel function $J_{2l}(z)$

$$J_{2l}(z) = \frac{1}{2\pi} \int_{-\pi}^{\pi} d\varphi \cos\left(z \sin\frac{\varphi}{2}\right) \cos l\varphi, \tag{23}$$

from which the derivatives and their integrals follow, we get for P_{1l} and P_{2l} the following formulae:

$$P_{il} = \frac{2G\omega}{r_i} \left(\left(E_i^2 (v_i^2 - 1) - \frac{m_i^4}{2E_i^2}\right) \int_0^{2v_i l} dx J_{2l}(x) \right.$$
$$\left. + 4E_i^2 (v_i^2 - 1) v_i^2 J'_{2l}(2v_i l) + 4E_i^2 v_i^4 J''_{2l}(2v_i l) \right), \quad i = 1, 2. \tag{24}$$

Using $r_2 = r_1 + \epsilon$, where ϵ is supposed to be small in comparison with radii r_1 and r_2, we obtain

$$\left[r_1^2 + r_2^2 + 2r_1 r_2 \cos\varphi\right]^{1/2} \approx 2a \cos\left(\frac{\varphi}{2}\right), \tag{25}$$

with

$$a = r_1\left(1 + \frac{\epsilon}{2r_1}\right). \tag{26}$$

So, instead of Eq. (19), we get

$$P_{12}(\omega, t) = \frac{2G\omega}{a\pi} \int_{-\infty}^{\infty} d\tau \frac{\sin\left[2\omega a \cos\left(\omega_0 \tau/2\right)\right]}{\cos\left(\omega_0 \tau\right)/2]} \cos\omega\tau$$
$$\times \left(E_1 E_2 \left(\omega_0^2 r_1 r_2 \cos\omega_0 \tau + 1\right)^2 - \frac{m_1^2 m_2^2}{2E_1 E_2} \right). \tag{27}$$

Now, we can approach the evaluation of the energy loss formula for the binary from the power spectral of Eqs. (24) and (27). The energy loss is defined by the relation

$$-\frac{dU}{dt} = \int P(\omega) d\omega =$$
$$\int d\omega \sum_{i,l} \delta(\omega - \omega_0 l) P_{il} + \int P_{12}(\omega) d\omega = -\frac{d}{dt}(U_1 + U_2 + U_{12}). \tag{28}$$

From [34] we have Kapteyn's formula:

$$\sum_{l=1}^{\infty} \frac{J_{2l}(2lv)}{l^2} = \frac{v^2}{2}. \tag{29}$$

After differentiating the last relation with respect to v, we have

$$\sum_{l=1}^{\infty} lJ'''_{2l}(2lv) = 0. \tag{30}$$

From [34] we learn other Kapteyn's formulae:

$$\sum_{l=1}^{\infty} 2lJ'_{2l}(2lv) = \frac{v}{(1-v^2)^2}, \tag{31}$$

and

$$\sum_{l=1}^{\infty} l\int_0^{2lv} J_{2l}(x)dx = \frac{v^3}{3(1-v^2)^3}. \tag{32}$$

So, after the application of Eqs. (30), (31) and (32) to Eqs. (24) and (28), we get

$$-\frac{dU_i}{dt} = \frac{Gm_i^2 v_i^3 \omega_0}{3r_i\left(1-v_i^2\right)^3}\left[13v_i^2 - 15\right]. \tag{33}$$

Instead of using Kapteyn's formulae for the interference term, we will perform a direct evaluation of the energy loss of the interference term by the ω-integration in (27) [35]. So, after some elementary modification in the ω-integral, we get

$$-\frac{dU_{12}}{dt} = \int_0^{\infty} P(\omega)d\omega =$$
$$A\int_{-\infty}^{\infty} d\tau \int_{-\infty}^{\infty} d\omega\omega e^{-i\omega\tau} \sin\left[2\omega a\cos\omega_0\tau\right]\left[\frac{B(C\cos\omega_0\tau+1)^2 - D}{\cos\left(\omega_0\tau/2\right)}\right], \tag{34}$$

with

$$A = \frac{G}{a\pi}, \quad B = E_1E_2, \quad C = v_1v_2, \quad D = \frac{m_1^2 m_2^2}{2E_1E_2}. \tag{35}$$

Using the definition of the δ-function and its derivative, we have, instead of Eq. (34), with $v = a\omega_0$

$$-\frac{dU_{12}}{dt} = A\omega_0\pi\int_{-\infty}^{\infty} dx\,\frac{\left[B(C\cos x+1)^2 - D\right]}{\cos\left(x/2\right)}$$
$$\times\left[\delta'(x - 2v\cos\left(x/2\right)) - \delta'(x + 2v\cos\left(x/2\right))\right]. \tag{36}$$

According to the Schwinger article [36], we express the delta function as follows:

$$\delta(x \pm 2v\cos\left(x/2\right)) = \sum_{n=0}^{\infty}\frac{\left(\pm 2v\cos\left(x/2\right)\right)^n}{n!}\left(\frac{d}{dx}\right)^n\delta(x). \tag{37}$$

Then

$$\delta'(x \pm 2v\cos\left(x/2\right)) = \sum_{n=0}^{\infty}\frac{\left(\pm 2v\cos\left(x/2\right)\right)^n}{n!}\left(\frac{d}{dx}\right)^{n+1}\delta(x) = \tag{38}$$

and it means that

$$\frac{[\delta'(x+2v\cos(x/2)) - \delta'(x-2v\cos(x/2))]}{\cos(x/2)} = (-2)\sum_{n=1}^{\infty}\frac{(2v)^{2n-1}(\cos(x/2))^{2(n-1)}}{(2n-1)!}\left(\frac{d}{dx}\right)^{2n}\delta(x) \tag{39}$$

Now, we can write Eq. (36) in the following form after some elementary operations:

$$-\frac{dU_{12}}{dt} = A\omega_0\pi\int_{-\infty}^{\infty} dx\left(B(C\cos x+1)^2 - D\right) \times(-2)\sum_{n=1}^{\infty}\frac{(2v)^{2n-1}(\cos(x/2))^{2(n-1)}}{(2n-1)!}\left(\frac{d}{dx}\right)^{2n}\delta(x), \tag{40}$$

where $\left(B(C\cos x+1)^2 - D\right)$ can be written as follows:

$$\left(B(C\cos x+1)^2 - D\right) = 4BC^2(\cos^4(x/2) + \left[4CB - 4BC^2\right](\cos^2(x/2) + \left[BC^2 - 2CB + B - D\right]. \tag{41}$$

After the application of the per partes method, we get from Eq. (40) the following mathematical object:

$$\begin{aligned}-\frac{dU_{12}}{dt} &= (-2)A\left[4BC^2\right]\omega_0\pi\int_{-\infty}^{\infty} dx\,\delta(x)\sum_{n=1}^{\infty}\left(\frac{d}{dx}\right)^{2n}(2v)^{2n-1}\frac{(\cos(x/2))^{2n+2}}{(2n-1)!}\\ &- 2A\left[4CB - 4BC^2\right]\omega_0\pi\int_{-\infty}^{\infty} dx\,\delta(x)\sum_{n=1}^{\infty}\left(\frac{d}{dx}\right)^{2n}(2v)^{2n-1}\frac{(\cos(x/2))^{2n}}{(2n-1)!}\\ &- 2A\left[BC^2 - 2CB + B - D\right]\omega_0\pi\int_{-\infty}^{\infty} dx\,\delta(x)\sum_{n=1}^{\infty}\left(\frac{d}{dx}\right)^{2n}(2v)^{2n-1}\frac{(\cos(x/2))^{2(n-1)}}{(2n-1)!}.\end{aligned} \tag{42}$$

We get after some elementary operations $\int\delta f(x) = f(0)$

$$J_1 = \sum_{n=1}^{\infty}\left(\frac{d}{dx}\right)^{2n}(2v)^{2n-1}\frac{(\cos(x/2))^{2n+2}}{(2n-1)!}\Bigg|_{x=0} = \sum_{n=0}^{\infty} f(n)v^{2n} = F(v^2), \tag{43}$$

$$J_2 = \sum_{n=1}^{\infty}\left(\frac{d}{dx}\right)^{2n}(2v)^{2n-1}\frac{(\cos(x/2))^{2n}}{(2n-1)!}\Bigg|_{x=0} = \sum_{n=0}^{\infty} g(n)v^{2n} = G(v^2) \tag{44}$$

and

$$J_3 = \sum_{n=1}^{\infty}\left(\frac{d}{dx}\right)^{2n}(2v)^{2n-1}\frac{(\cos(x/2))^{2(n-1)}}{(2n-1)!}\Bigg|_{x=0} = \sum_{n=0}^{\infty} h(n)v^{2n} = H(v^2) \tag{45}$$

where f, g, h, F, G, H are functions which must be determined.
So we get instead of Eq. (41) the following final form:

$$\begin{aligned}-\frac{dU_{12}}{dt} &= (-2)A\left[4BC^2\right]\omega_0\pi G(v^2) - 2A\left[4CB - 4BC^2\right]\omega_0\pi F(v^2)\\ &- 2A\left[-2CB + BC^2 + B - D\right]\omega_0\pi H(v^2)\end{aligned} \tag{46}$$

Let us remark that we can use simple approximation in Eq. (41) as follows: $(\cos(x/2))^{2n+2} \approx (\cos(x/2))^2$, $(\cos(x/2))^{2n} \approx (\cos(x/2))^2$, $(\cos(x/2))^{2(n-1)} \approx (\cos(x/2))^2$. Then, after using the well-known formula

$$\left(\frac{d}{dx}\right)^{2n} \cos^2(x/2) = \frac{1}{2}\cos(x+\pi n) \tag{47}$$

and

$$\frac{1}{2}\cos(x+\pi n)\bigg|_{x=0} = \frac{1}{2}(-1)^n. \tag{48}$$

So, instead of Eq. (46), we have

$$-\frac{dU_{12}}{dt} = A\omega_0\pi\{2BC + BC^2 + B - D\}\sum_{n=1}^{\infty}\frac{(2v)^{2n-1}(-1)^n}{(2n-1)!}. \tag{49}$$

2. Energy shift of H-atom electrons due to the black hole thermal bath

We here determine the electromagnetic shift of energy levels of H-atom electrons by calculating an electron coupling to the black hole thermal bath. The energy shift of electrons in H-atom is determined in the framework of nonrelativistic quantum mechanics.

The Gibbons-Hawking effect is the statement that a temperature can be associated to each solution of the Einstein field equations that contain a causal horizon.

Schwarzschild space-time involves an event horizon associated with temperature T of a black hole of mass M. We consider here the influence of the heat bath of the Gibbons-Hawking photons on the energy shift of H-atom electrons.

The analogical problems are solved in the scientific respected journals. There is a general conviction of an analogy between the black hole and the hydrogen atom. Corda [37] used the model where Hawking radiation is a tunneling process. In his article the emission is expressed in terms of the black hole quantum levels. So, the Hawking radiation and black hole quasinormal modes by Corda [38] are analogical to hydrogen atom by Bohr.

In this model [39] the corresponding wave function is written in terms of a unitary evolution matrix. So, the final state is a pure quantum state with no information loss. Black hole is defined as the quantum systems, with discrete quantum spectra, with Hooft's assumption that Schrödinger equations are universal for all universe dynamics.

Thermal photons by Gibbons and Hawking are blackbody photons, with the Planck photon distribution law [40–42], derived from the statistics of the oscillators inside of the blackbody. Later Einstein [43] derived the Planck formula from the Bohr model of atom where photons and electrons have the discrete energies related with the Bohr formula $\hbar\omega = E_i - E_f$, E_i, E_f being the initial and final energies of electrons.

Now, we determine the modification of the Coulomb potential due to blackbody photons. At the start, the energy shift in the H-atom is the potential $V_0(\mathbf{x})$, generated by nucleus of the H-atom. The potential at point $V_0(\mathbf{x} + \delta\mathbf{x})$ is [44, 45]

$$V_0(\mathbf{x} + \delta\mathbf{x}) = \left\{1 + \delta\mathbf{x}\nabla + \frac{1}{2}(\delta\mathbf{x}\nabla)^2 + \ldots\right\}V_0(\mathbf{x}). \tag{50}$$

The average of the last equation in space enables the elimination of the so-called the effective potential:

$$V(\mathbf{x}) = \left\{1 + \frac{1}{6}(\delta\mathbf{x})_T^2\Delta + \dots\right\}V_0(\mathbf{x}), \tag{51}$$

where $(\delta\mathbf{x})_T^2$ is the average value of the square coordinate shift caused by the thermal photons. The potential shift follows from Eq. (51):

$$\delta V(\mathbf{x}) = \frac{1}{6}(\delta\mathbf{x})_T^2\Delta V_0(\mathbf{x}). \tag{52}$$

The shift of the energy levels is given by the standard quantum formula [44]:

$$\delta E_n = \frac{1}{6}(\delta\mathbf{x})_T^2(\psi_n\Delta V_0\psi_n). \tag{53}$$

In case of the Coulomb potential, which is the case of the H-atom, we have

$$V_0 = -\frac{e^2}{4\pi|\mathbf{x}|}. \tag{54}$$

Then for the H-atom we can write

$$\delta E_n = \frac{2\pi}{3}(\delta\mathbf{x})_T^2\frac{e^2}{4\pi}|\psi_n(0)|^2, \tag{55}$$

where we used the following equation for the Coulomb potential

$$\Delta\frac{1}{|\mathbf{x}|} = -4\pi\delta(\mathbf{x}). \tag{56}$$

The motion of electron in the electric field is evidently described by elementary equation:

$$\delta\ddot{\mathbf{x}} = \frac{e}{m}\mathbf{E}_T, \tag{57}$$

which can be transformed by the Fourier transformation into the following equation

$$|\delta\mathbf{x}_{T\omega}|^2 = \frac{1}{2}\left(\frac{e^2}{m^2\omega^4}\right)\mathbf{E}_{T\omega}^2, \tag{58}$$

where the index ω concerns the Fourier component of the above functions.
Using Bethe idea [46] of the influence of vacuum fluctuations on the energy shift of electron, the following elementary relations were applied by Welton [45], Akhiezer et al. [44] and Berestetzkii et al. [47]:

$$\frac{1}{2}\mathbf{E}_\omega^2 = \frac{\hbar\omega}{2}, \tag{59}$$

and in case of the thermal bath of the blackbody, the last equation is of the following form [48]:

$$\mathbf{E}_{T\omega}^2 = \varrho(\omega) = \left(\frac{\hbar\omega^3}{\pi^2c^3}\right)\frac{1}{e^{\frac{\hbar\omega}{kT}} - 1}, \tag{60}$$

because the Planck law in (60) was written as

$$\varrho(\omega) = G(\omega) < E_\omega > = \left(\frac{\omega^2}{\pi^2 c^3}\right) \frac{\hbar\omega}{e^{\frac{\hbar\omega}{kT}} - 1}, \tag{61}$$

where the term

$$< E_\omega > = \frac{\hbar\omega}{e^{\frac{\hbar\omega}{kT}} - 1} \tag{62}$$

is the average energy of photons in the blackbody and

$$G(\omega) = \frac{\omega^2}{\pi^2 c^3} \tag{63}$$

is the number of electromagnetic modes in the interval ω, $\omega + d\omega$.
Then,

$$(\delta \mathbf{x}_{T\omega})^2 = \frac{1}{2}\left(\frac{e^2}{m^2\omega^4}\right)\left(\frac{\hbar\omega^3}{\pi^2 c^3}\right)\frac{1}{e^{\frac{\hbar\omega}{kT}} - 1}, \tag{64}$$

where $(\delta \mathbf{x}_{T\omega})^2$ involves the number of frequencies in the interval $(\omega, \omega + d\omega)$.
So, after some integration, we get

$$(\delta \mathbf{x})_T^2 = \int_{\omega_1}^{\omega_2} \frac{1}{2}\left(\frac{e^2}{m^2\omega^4}\right)\left(\frac{\hbar\omega^3}{\pi^2 c^3}\right)\frac{d\omega}{e^{\frac{\hbar\omega}{kT}} - 1} = \frac{1}{2}\left(\frac{e^2}{m^2}\right)\left(\frac{\hbar}{\pi^2 c^3}\right)F(\omega_2 - \omega_1), \tag{65}$$

where $F(\omega)$ is the primitive function of the omega-integral with

$$\frac{1}{\omega}\frac{1}{e^{\frac{\hbar\omega}{kT}} - 1}, \tag{66}$$

which is not elementary, and it is not in the tables of integrals.

Frequencies ω_1 and ω_2 can be determined from the field of thermal photons. It was performed for the Lamb shift [44, 47] caused by the interaction of the Coulombic atom with the field fluctuations. The Bethe-Welton method is valid here too and so we take Bethe-Welton frequencies. It means an electron does not respond to the fluctuating field if the frequency is much less than the atom binding energy given by the Rydberg constant [49] $E_{Rydberg} = \alpha^2 mc^2/2$. So, the lower frequency limit is

$$\omega_1 = E_{Rydberg}/\hbar = \frac{\alpha^2 mc^2}{2\hbar}, \tag{67}$$

where $\alpha \approx 1/137$ is so-called the fine structure constant.

The second frequency follows from the cutoff, determined by the neglection of the relativistic effect in our theory. So, we write

$$\omega_2 = \frac{mc^2}{\hbar}. \tag{68}$$

If we express the thermal function in the form of the geometric series

$$\frac{1}{e^{\frac{\hbar\omega}{kT}} - 1} = q(1 + q^2 + q^3 + ...); \quad q = e^{-\frac{\hbar\omega}{kT}}, \tag{69}$$

$$\int_{\omega_1}^{\omega_2} q\left(1+q^2+q^3+\ldots..\right)\frac{1}{\omega}d\omega = \ln|\omega| + \sum_{k=1}^{\infty}\frac{\left(-\frac{\hbar\omega}{kT}\right)^k}{k!k} + \ldots.; \quad q = e^{-\frac{\hbar\omega}{kT}} \tag{70}$$

and the first thermal contribution is

$$Thermal\ \ contribution = \ln\frac{\omega_2}{\omega_1} - \frac{\hbar}{kT}(\omega_2 - \omega_1), \tag{71}$$

then, with Eq. (55)

$$\delta E_n \approx \frac{2\pi}{3}\left(\frac{e^2}{m^2}\right)\left(\frac{\hbar}{\pi^2 c^3}\right)\left(\ln\frac{\omega_2}{\omega_1} - \frac{\hbar}{kT}(\omega_2 - \omega_1)\right)|\psi_n(0)|^2, \tag{72}$$

where according to Sokolov et al. [50]

$$|\psi_n(0)|^2 = \frac{1}{\pi n^2 a_0^2} \tag{73}$$

with

$$a_0 = \frac{\hbar^2}{me^2}. \tag{74}$$

Let us only remark that the numerical form of Eq. (72) has deep experimental astrophysical meaning.

Haroche [51] and his group performed experiments with the Rydberg atoms in a cavity. We used here Gibbons-Hawking black hole for the determination of the energy shift of H-atom electrons in the black hole gas.

3. Velocity of sound in the black hole photon gas

We have seen that the black hole can be modeled by the blackbody, and it means that there is the velocity of sound in the Gibbons-Hawking black hole thermal bath. So, let us derive the sound velocity from the thermodynamics of photon gas and energy mass relation.

In order to be pedagogically clear, we start with the derivation of the speed of sound in the real elastic rod.

Let A be the cross-section of the element Adx of a rod on the axis x. Let $\varphi(x,t)$ be the deflection of Adx at point x at time t. The shift of the Adx at point $x+dx$ is evidently

$$\varphi + \frac{\partial\varphi}{\partial x}dx. \tag{75}$$

Now, we suppose that the force tension $F(x,t)$ acting on the Adx of the rod is given by Hooke's law:

$$F(x,t) = EA\frac{\partial\varphi}{\partial x}, \tag{76}$$

where E is Young's modulus of elasticity. We easily derive that

$$F(x+dx) - F(x) \approx EA\frac{\partial^2\varphi}{\partial x^2}dx. \tag{77}$$

The mass of Adx is ϱAdx, where ϱ is the mass density of the rod and the dynamical equilibrium is expressed by Newton's law of force:

$$\varrho Adx\varphi_{tt} = EA\varphi_{xx}dx \tag{78}$$

or

$$\varphi_{tt} - v^2\varphi_{xx} = 0, \tag{79}$$

where

$$v = \left(\frac{E}{\varrho}\right)^{1/2} \tag{80}$$

is the velocity of sound in the rod.

The complete solution of Eq. (79) includes the initial and boundary conditions. We suppose that Eq. (80) is of the universal validity also for gas in the cylinder tube. If $(\Delta L/L)$ is the relative prolongation of a rod, then an analogue for the tube of gas is $\Delta V/V$, $F \to \Delta p$, where V is the volume of a gas and p is gas pressure. Then, the modulus of elasticity as the analogue of Eq. (76) is

$$E = -\frac{dp}{dV}V. \tag{81}$$

The sound in ideal gas is the adiabatic thermodynamic process with no heat exchange. This is the model of the sound spreading in the gas of blackbody photons. Such process is described by the thermodynamic equation:

$$pV^\kappa = const, \tag{82}$$

where κ is the Poisson constant defined as $\kappa = c_p/c_v$, with c_p, c_v being the specific heat under constant pressure and under constant volume.

After differentiation of Eq. (82), we get the following equation:

$$dpV^\kappa + \kappa V^{\kappa-1}dV = 0, \tag{83}$$

or

$$\frac{dp}{dV} = -\kappa\frac{p}{V}. \tag{84}$$

After inserting Eq. (84) into Eq. (81), we get from Eq. (80) the so-called Newton-Laplace formula:

$$v = \sqrt{\kappa\frac{p}{\varrho}}, \tag{85}$$

with ϱ being the gas mass density.

The equilibrium radiation density has the Stefan-Boltzmann form:

$$u = aT^4; \quad a = 7,5657.10^{-16}\frac{\mathrm{J}}{\mathrm{K}^4\mathrm{m}^3}. \tag{86}$$

Then, with regard to the thermodynamic definition of the specific heat,

$$c_v = \left(\frac{\partial u}{\partial T}\right)_V = 4aT^3. \tag{87}$$

Similarly, with regard to the general thermodynamic theory,

$$c_p = c_v + \left[\left(\frac{\partial u}{\partial V}\right)_T + p\right]\left(\frac{\partial V}{\partial T}\right)_p = c_v, \tag{88}$$

because $\left(\frac{\partial V}{\partial T}\right)_T = 0$ for photon gas, and in such a way, $\kappa = 1$ for photon gas. According to the theory of relativity, there is a relation for mass and energy, namely, $m = E/c^2$. At the same time, the pressure and the internal energy of the blackbody gas are related as $p = u/3$. So, in our case

$$\varrho = u/c^2 = \frac{aT^4}{c^2}; \quad p = \frac{u}{3}. \tag{89}$$

So, after the insertion of formulae in Eq. (88) into Eq. (85), the final formula for the sound velocity in photon blackbody sea is the following:

$$v = c\sqrt{\frac{\kappa}{3}} = \frac{c}{3}\sqrt{3}, \tag{90}$$

which was derived by Partovi [52] using the QED theory of the photon gas. We correctly derived $v/c < 1$.

So, we have performed the derivation of the velocity of sound in the relic photon sea. It is not excluded that the relic sound can be detected by the special microphones of Bell Laboratories. If we use van der Waals equation of state or the Kamerlingh Onnes virial equation, the obtained results will be modified with regard to the basic results.

Our derivation of the light velocity in the blackbody photon gas was based on the classical thermodynamic model with the adiabatic process ($\delta Q = 0$), controlling the spreading of sound in the gas. Partovi [52] derived additional radiation corrections to the Planck distribution formula and the additional correction to the speed of sound in the relic photon sea. His formula is of the form

$$v_{sound} = \left[1 - \frac{88\pi^2\alpha^2}{2025}\left(\frac{T}{T_e}\right)^4\right]\frac{c}{\sqrt{3}}, \tag{91}$$

where α is the fine structure constant and $T_e = 5.9$ G Kelvin. We see that our formula is the first approximation in the Partovi expression.

There is the Boltzmann statistical theory of transport of sound energy in a gas [53]. After the application of this theory to the photon gas or relic photon gas, we can obtain results involving the cross-section of the photon-photon interaction [47]:

$$\sigma_{\gamma\gamma} = 4,7\alpha^4\left(\frac{c}{\omega}\right)^2; \quad \hbar\omega \ll mc^2, \tag{92}$$

and

$$\sigma_{\gamma\gamma} = \frac{973}{10125\pi}\alpha^2 r_e^2\left(\frac{\hbar\omega}{mc^2}\right)^6; \quad \hbar\omega \gg mc^2, \tag{93}$$

where $r_e = e^2/mc^2 = 2,818 \times 10^{-13}$ cm is the classical radius of electron and $\alpha = e^2/\hbar c$ is the fine structure constant with numerical value $1/\alpha = 137,04$.

4. Discussion and summary

We have derived the spectral density of gravitons and the total quantum loss of energy of the black hole binary. The energy loss is caused by the emission of gravitons during the motion of the two black hole binaries around each other under their gravitational interaction. The energy loss formulae of the production of gravitons are derived here by the Schwinger method. Because the general relativity and theory of gravity do not necessarily contain the last valid words to be written about the nature of gravity and it is not, of course, a quantum theory [21], they cannot give the answer on the production of gravitons and the quantum energy loss, respectively. So, this article is the original text that discusses the quantum energy loss caused by the production of gravitons by the black hole binary system. It is evident that the production of gravitons by the binary system forms a specific physical situation, where a general relativity can be seriously confronted with the source theory of gravity.

This article is an extended version of an older article by the present author [33], in which only the spectral formulae were derived. Here we have derived the quantum energy loss formulae, with no specific assumption concerning the strength of the gravitational field. We hope that future astrophysical observations will confirm the quantum version of the energy loss of the binary black hole.

In the next part of the chapter, the electromagnetic shift of energy levels of H-atom electrons was determined by calculating an electron coupling to the Gibbons-Hawking electromagnetic field thermal bath of the black hole. The energy shift of electrons in H-atom is determined in the framework of nonrelativistic quantum mechanics.

In the last section, we have determined the velocity of sound in the blackbody gas of photons inside of the black hole. Derivation was based on the thermodynamic theory of the photon gas and the Einstein relation between energy and mass. The spectral form for the n-dimensional blackbody was not here considered. The text is based mainly on the author articles published in the international journals of physics [33, 54, 55].

There is the fundamental problem concerning the maximal mass of the black hole. The theory of the space–time with maximal acceleration constant was derived by authors [56, 57]. In this theory the maximal acceleration constant is the analogue of the maximal velocity in special theory of relativity. Maximal acceleration determines the maximal black hole mass where the mass of the black hole is restricted by maximal acceleration of a body falling in the gravity field of the black hole.

Another question is what is the relation of our formulae to the results obtained by LIGO (Laser Interferometer Gravitational-Wave Observatory)? LIGO is the largest and most sensitive interferometer facility ever built. It has been periodically upgraded to increase its sensitivity. The most recent upgrade, Advanced LIGO (2015), detected for the first time the gravitational wave, with sensitivity far above the background noise. The event with number GW150914 was identified with the result of a merger of two black holes at a distance of about 400 Mpc from Earth [58]. Two additional significant detections, GW151226 and GW170104, were reported later. We can say that at this time it is not clear if the LIGO results involve information on the spectrum of gravitons calculated in this chapter.

DOI: http://dx.doi.org/10.5772/intechopen.82659

Author details

Miroslav Pardy
Department of Physical Electronics, Masaryk University, Brno, Czech Republic

*Address all correspondence to: pamir@physics.muni.cz

References

[1] Schwarzschild K. Über das Gravitationsfeld eines Massenpunktes nach der Einsteinschen Theorie. Sitsungsberichte Königlich Preussischen Akademie der Wissenschaften. 1916;**1**: 189. [English translation: On the gravitational field of a mass point according to Einstein's theory. General Relativity and Gravitation. 2003;**35**:951]

[2] Finkelstein D. Past-future asymmetry of the gravitational field of a point particle. Physics Review. 1958;**110**:965

[3] Kruskal MD. Maximal extension of Schwarzschild metric. Physics Review. 1960;**119**:1743

[4] Kerr RP. Gravitational field of a spinning mass as an example of algebraically special metrics. Physical Review Letters. 1963;**11**:237

[5] Vishveshwara CV. Scattering of gravitational radiation by a Schwarzschild black hole. Nature (London). 1970;**227**:936

[6] Press WH. Long wave-trains of gravitational waves from a vibrating black hole. The Astrophysical Journal. 1971;**170**:L105

[7] Chandrasekhar S, Detweiler SL. The quasi-normal modes of the Schwarzschild black hole. Proceedings of the Royal Society A. 1975;**344**:441

[8] Blanchet L, Damour LT, Iyer BR, Will CM, Wiseman AG. Gravitational-radiation damping of compact binary systems to second-post-Newtonian order. Physical Review Letters. 1995; **74**:3515

[9] Blanchet L. Gravitational radiation from post-Newtonian sources and inspiralling compact binaries. Living Reviews in Relativity. 2014;**17**:2

[10] Buonanno A, Damour T. Effective one-body approach to general relativistic two-body dynamics. Physical Review D. 1999;**59**:084006

[11] Pretorius F. Evolution of binary black hole spacetimes. Physical Review Letters. 2005;**95**:121101

[12] Campanelli M, Lousto CO, Marronetti P, Zlochower Y. Accurate evolutions of orbiting black-hole binaries without excision. Physical Review Letters. 2006;**96**:111101

[13] Baker JG, Centrella J, Choi DI, Koppitz M, van Meter J. Gravitational-wave extraction from an inspiraling configuration of merging black holes. Physical Review Letters. 2006;**96**: 111102

[14] Webster BL, Murdin P. Is Cygnus X-1-a spectroscopic binary with a heavy companion? Nature (London). 1972;**235**:37

[15] Bolton CT. Is Cyg X-1 a neutron star? Nature (London). 1972;**240**:124

[16] Casares J, Jonker PG. Mass measurements of stellar and intermediate-mass black holes. Space Science Reviews. 2014;**183**:223

[17] Hulse RA, Taylor JH. Discovery of a pulsar in a binary system. The Astrophysical Journal. 1975;**195**:L51

[18] Taylor JH, Weisberg JM. A new test of general relativity: Gravitational radiation and the binary pulsar PSR B**1913+16**. The Astrophysical Journal. 1982;**253**:908

[19] Aasi J et al. Advanced Virgo: A second-generation interferometric gravitational wave detector. Classical and Quantum Gravity. 2015;**32**:074001

[20] Acernese F et al. Advanced Virgo: A 2nd generation interferometric gravitational wave detector. Classical and Quantum Gravity. 2015;**32**:024001

[21] Taylor JH Jr. Binary pulsars and relativistic gravity. Nobel Lecture. December 1993. www.nobelprise.org/uploads/2018/06/taylor-lecture-1.pdf

[22] Gold T. Rotating neutron stars as the origin of the pulsating radio sources nature. Nature. 1968;**218**:731

[23] Goldreich P, Julian WH. Pulsar electrodynamics. The Astrophysical Journal. 1969;**157**:869

[24] Damour T, Taylor JH. Strong field tests of relativistic gravity and binary pulsars. Physical Review D. 1992;**45**(6): 1840-1868

[25] Taylor JH, Wolszczan A, Damour T, Weisberg JM. Experimental constrains on strong-field relativistic gravity. Nature. 1999;**355**:132-136

[26] Manoukian EB. A quantum viewpoint of gravitational radiation. General Relativity and Gravitation. 1990;**22**:501

[27] Schwinger J, Tsai WY, Erber T. Classical and quantum theory of synergic synchrotron—Cherenkov radiation. Annals of Physics (New York). 1976;**96**:303

[28] Dittrich W. Source methods in quantum field theory. Fortschritte der Physik. 1978;**26**:289

[29] Schwinger J. Gravitons and photons: The methodological unification of source theory. General Relativity and Gravitation. 1976;**7**:251

[30] Schwinger J. Particles, Sources and Fields. Vol. 1. Reading, MA: Addison-Wesley Publ. Comp. 1970

[31] Weinberg S. Gravitation and Cosmology. New York: John Wiley and Sons, Inc.; 1972

[32] Cho CF, Dass NDH. Gravitational two-body problems, a source theory viewpoint. Annals of Physics (New York). 1976;**90**:406

[33] Pardy M. The Synchrotron production of gravitons by the binary system. General Relativity and Gravitation. 1983;**15**(11):1027

[34] Prudnikov AP, Bryčkov JA, Maričev OI. Integrals and Series—Special Functions. Moscow: Nauka; 1983 (in Russian)

[35] Sokolov AA, Ternov IM. The Relativistic Electron. Moscow: Nauka; 1983 (in Russian)

[36] Schwinger J. On radiation by Electrons in a Betatron. LNBL-39088, CBP Note-179, UC-414; Berkeley, CA; January 13, 1998

[37] Corda C. Precise model of Hawking radiation from the tunneling mechanism. Classical and Quantum Gravity. 2015;**32**:195007

[38] Corda CH. Quasi-normal modes: The "electrons" of black holes as "gravitational atoms"? Implications for the black hole information puzzle. Advances in High Energy Physics. 2015. Artical ID: 867601. http://dx.doi.org/10.1155/2015/867601

[39] Corda C. Time dependent Schrödinger equation for black hole evaporation: No information loss. Annals of Physics. 2015;**353**:71

[40] Planck M. Zur theorie des gesetzes der energieverteilung im normalspektrum. Verhandlungen der Deutschen Physikalischen Gesellschaft. 1900;**2**:237

[41] Planck M. Ueber irreversible strahlungsvorgänge. Ann. Phys. 1901;**4**: 553

[42] Schöpf H-G. Theorie der Wärmestrahlung in Historisch—Kritischer Darstellung. Berlin: Alademie/Verlag; 1978

[43] Einstein A. Zur quantentheorie der strahlung. Physikalische Zeitschrift. 1917;**18**:121

[44] Akhiezer AI, Berestetzkii VB. Quantum Electrodynamics. Moscow: GITTL; 1953

[45] Welton T. Some observable effects of the quantum-mechanical fluctuations of the electromagnetic field. Physics Review. 1948;**74**:1157

[46] Bethe HA. The electromagnetic shift of energy levels. Physics Review. 1947;**2**:339

[47] Berestetzkii VB, Lifshitz EM, Pitaevskii LP. Quantum Electrodynamics. Oxford: Butterworth-Heinemann; 1999

[48] Isihara A. Statistical mechanics. New York/London: Academic Press; 1971

[49] Rohlf JW. Modern physics from $\alpha\ to\ Z^0$. London/New York: John Wiley & Sons Ltd.; 1994

[50] Sokolov AA, Loskutov YM, Ternov IM. Quantum Mechanics (State Pedagogical Edition). Moscow: Holt, Rinehart and Wilson Unstated Edition; 1966 (in Russian)

[51] Haroche S. The secrets of my prizewinning research. Nature. 2012; **490**:311

[52] Partovi MH. QED corrections to Planck's radiation law and photon thermodynamics. Physical Review D. 1994;**50**(2):1118-1124

[53] Uhlenbeck GE, Ford GW. Lectures in Statistical Physics. Providence, RI: American Mathematical Society; 1963

[54] Pardy M. Energy shift of H-atom electrons due to Gibbons-Hawking thermal bath. Journal of High Energy Physics, Gravitation and Cosmology. 2016;**2**:472-477

[55] Pardy M. Velocity of sound in the black-body. Results in Physics. 2013;**3**: 7073

[56] Pardy M. Thomas precession by uniform acceleration and gravity. Journal of High Energy Physics, Gravitation and Cosmology. 2018; **2018**(4):132-142

[57] Pardy M. From Lobachevsky-Beltrami-Fok Space to Schwinger Gravity and Cosmology. Saarbrücken, Detschland/Germany: LAMBERT Academic Publishing; 2017

[58] LIGO. The Laser Interferometer Gravitational-Wave Observatory. The LIGO Scientific Collaboration. 2015. Available from: http://www.ligo.org

Chapter 5

Applications of the Abelian Vortex Model to Cosmic Strings and the Universe Evolution

Mikael Souto Maior de Sousa and Anderson Alves de Lima

Abstract

Due to the wide range of applications and effects of the Abelian vortex model of Nielsen and Olesen in the many areas of physics, ranging from condensed matter to astrophysical effects, some work in the literature is necessary to approach this topic in a succinct form that the undergraduate student in both physics and related areas has the possibility to know and understand. The mechanisms associated with this vortex model indicate him as a strong candidate for the source for the topological defects proposed by Vilenkin.

Keywords: cosmic string, curved space-time, relativity, field theory

1. Introduction

According to the Big Bang theory, the universe is expanding and cooling. During its expansion, the spontaneous breaks of fundamental symmetries led the universe to undergo a series of phase transitions. In high-energy physics models, the formation of topological defects, caused by transitions, such as domain walls, monopoles, and cosmic strings, among others, is predicted to occur according to the reference [1, 2].

The cosmic string is among the most studied types of topological defects, although recent observations of cosmic background radiation have discarded it as the primary source for primordial density perturbations. Such a defect still serves as one of the contributions of this disturbance. This type of defect also serves as a possible source for explaining a considerable number of astrophysical effects, such as: bursts of gamma rays, where the energy scale of the string in which the symmetry is broken, on an energy scale of the order of 10^{14} GeV, explains the rate, duration, and fluency of gamma ray bursts [3]; high-frequency gravitational wave emissions, which have as a consequence of these emissions the stochastic set of gravitational waves generated by a cosmological network of non-Gaussian loops [4]; and the generation of high-energy cosmic rays [5]. The cosmic rays of high-energy particles may have originated during the process of collapse and/or annihilation of topological defects associated with the great unification theories.

In condensed matter physics, it is well known that superconductors almost completely exclude any external magnetic field if it is less than a critical value (Meissner effect) [6]. However, for type 2 superconducting, which are formed by materials in which the transition to the superconducting state is gradual, in the presence of an intermediate state, if the external field is increased to a certain value greater than the critical value, such field. This superconductor passes through a magnetic flux tube form. These phenomena are called magnetic flux vortices which, in turn, are quantized.

The possibility of the theoretical existence of such vortices was first demonstrated by Abrikosov [7]. He showed that these naturally occur as solutions to the Ginsburg-Landau theory of superconductivity in the presence of an external magnetic field. Following this theory, the existence of such objects was verified experimentally, and many of their properties were rigorously investigated in [6]. Some years later, Nielsen and Olesen [8] showed, starting from the relativistic field theory model with spontaneous break of symmetry, more specifically of the Abelian Higgs model interacting with a field of gauge, that this system presents solutions with cylindrical symmetry carrying a magnetic flux. These configurations correspond to vortex solutions.

The analysis of the influence of this system on space-time geometry was performed by Garfinkle [9] and Laguna [10]. In their works, the authors coupled the energy-momentum tensor, associated to the Nielsen-Olesen model, with the Einstein field equations. In this sense, they have shown that the vortex has an internal structure characterized by the nonzero magnetic flux that runs along it, the extent of which is determined by the energy scale at which the symmetry is broken. Two scale lengths appear naturally, one related to the extent of the magnetic flux which, in turn, is proportional to the inverse of the vector field mass, m_v, east field, which acquires mass due to the Higgs mechanism; and the other associated with the inverse of the scalar field mass, m_s, the latter, as a measure of the point where the scalar field decreases to its vacuum value. Moreover, the authors also analyzed the geometry of space-time and verified that asymptotically the surface perpendicular to the vortex corresponds to Minkowski's space-time minus a slice, resulting in a space with an angular deficit.

A special vortex solution satisfying the Bogomolny-Prasad-Sommerfield (BPS) boundary [11, 12] shows the masses of the scalar field and of the same caliber field, that is, $m_s = m_v$. For this case, Linet [13] was able to find an exact solution for the metric tensor, which is determined in terms of the energy density of the cosmic string. In this limit, the surface perpendicular to the line of the solution of vortex has a conical structure and, the space-time surrounding, corresponds to the space-time of an idealized cosmic string.

At great distances, the space-time generated by a cosmic string has, in its origin and in the orthogonal plane to the disposition of this object, a conical topology with a planar angle deficit proportional to the linear density of mass of this cosmic string. In quantum field theory, the nontrivial topology of this object induces non-vanish vacuum expected values for physical observables. These vacuum polarization effects can be interpreted as a modification in the quantum levels of the lower energy state of a theory. In quantum field theory, induced by a conic structure, they were the targets of many works published. For example, we can observe several published works, taking into account the case for scalar fields [14–19] and fermionic fields [20–22] interacting with vector fields. Another induced physical observable, due to the presence of this defect, is the current and charge density, which will serve as the source for Maxwell's equations. Such an object considering fermionic fields is seen in [23–26].

2. The general relativity and the space-time

The general relativity theory is a geometric theory of gravitation published by Albert Einstein in 1915 and the current description of gravitation in modern physics.

It is a set of hypotheses that generalizes Newton's special relativity and the universal gravitation law providing a unified description of gravity as a geometric property of the space-time. In particular, the "curvature of space-time" is directly related to the energy and moment of any matter and radiation present. The relation is specified by Einstein's field equations, a system of partial differential equations.

All geometric information about the space-time would be contained in this mathematical object called, formally, metric tensor, $g_{\mu\nu}$. In other words, the distribution of matter and energy tells how the geometry of space-time [27] must be. The equation proposed by Einstein for the theory of General Relativity is given by the expression below

$$R_{\mu\nu} - \frac{1}{2} g_{\mu\nu} R = 8\pi T_{\mu\nu}. \quad (1)$$

Here, $R_{\mu\nu}$ is the Ricci tensor that is obtained from the Riemann tensor, $R = g^{\mu\nu} R_{\mu\nu}$ is the scalar of curvature, and $T_{\mu\nu}$ is the energy-momentum tensor. In order to introduce the idea of the metric structure of the space-time, we will briefly review the necessary basic concepts, such as inertial frame and interval of events [27].

Let us suppose that an inertial frame S is described in Cartesian coordinates (t, x, y, z). In this frame, we have the line element ds being infinitesimal and having its own time interval (event) given by

$$ds^2 = dt^2 - dx^2 - dy^2 - dz^2. \quad (2)$$

But if we consider a non-inertial reference system, S', for example, the line element will not be given, in general, by the sum of the squares of the coordinate differentials. In this case, for a better understanding, let us consider an event in a rotating frame, around the z axis, whose angular frequency of rotation is ω. Let (t', x', y', z') be the coordinates of this new S' referential. The relation between both reference frames may be illustrated by **Figure 1**.

The general coordinate transformations between the both reference frames S and S′ are given as follow,

$$x = x' \cos(\omega t) - y' sen(\omega t) \quad (3)$$

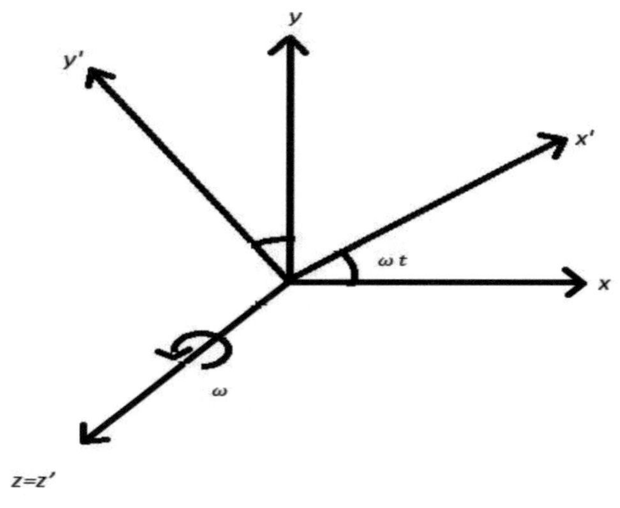

Figure 1.
The relation between S and S′ reference frame with angular velocity ω around the z = z′ axis.

$$y = x' sen(\omega t) + y' \cos(\omega t) \tag{4}$$

$$z = z'. \tag{5}$$

In this way, taking into account the derivative of the Eqs. (3)–(5) and putting them in the Eq. (2), the line element will take the form expressed by

$$ds^2 = \left[1 - \omega^2\left(x'^2 + y'^2\right)\right]dt^2 + 2\omega dt(y'dx' - x'dy') - d\vec{r}\ '2. \tag{6}$$

We see, therefore, that the line element is not only the sum or difference of the squares of the differential coordinates.

Looking to the Eq. (2), we identify that $ds^2 = \eta_{\mu\nu}dx^\mu dx^\nu$, where we have the metric signature given by $\eta_{\mu\nu} = (1, -1, -1, -1)$ being the four-vector position $x^\mu = \left(t, -\vec{r}\right)$. On the other hand, looking into Eq. (6), when non-inertial coordinate systems are used, the line element will include terms that are products of the different coordinate differentials. So, we can write the line element as follows

$$ds^2 = g_{\mu\nu}(x)dx^\mu dx^\nu. \tag{7}$$

Now, $g_{\mu\nu}(x)$ represents a set of ten functions of the space and time coordinates and it is symmetric, i.e., $g_{\mu\nu}(x) = g_{\nu\mu}(x)$. The system described by Eq. (7) is called "curved system" and corresponds to an accelerated reference system. The functions $g_{\mu\nu}(x)$ contain all the geometric properties of the space-time. For the case where we deal with inertial frames, we just have $g_{\mu\nu}(x) = \eta_{\mu\nu}$.

Einstein showed that accelerated referential are equivalent to gravitational fields so that gravitational effects will be described by the metric tensor, $g_{\mu\nu}(x)$. In this case, the gravitation can be understood as a deviation in the metric of the space-time plane. Moreover, this metric is not fixed arbitrarily but will depend on the local distribution of matter.

In fact, this equivalence is verified only locally. In a non-inertial system, given a metric $g_{\mu\nu}(x)$, we can always reduce it globally to the Galileo form, Eq. (2), by means of a suitable coordinate transformation. On the other hand, a gravitational field cannot be eliminated globally by a coordinate transformation, and the metric can only be reduced to the flat form (Minkowski) only in a very small finite region of the space, i.e., locally. When such a situation occurs, the space-time is called pseudo-Riemannian space-time.

3. Cosmic strings

It is believed that fluctuations that gave rise to the large-scale structures of the Universe must have a primordial origin, that is, they are associated with the first moments after the Big Bang. The existing theories for structure formation in the Universe fall into two categories.

One of them based on amplification of quantum fluctuations in a scalar field during inflation. The other one based on a phase transition with symmetry breaking in the primordial universe that gives rise to the formation of topological defects.

Seen from the moment of creation, the Universe goes through phase successions. The transitions between the first of these phases occur when the Universe is dominated by a quantum gravitation whose exact contours are unknown but during which the interactions are thought to be unified and characterized by a high degree of symmetry. These transitions imply symmetry breaks and can have important implications including the formation of topological defects such as the formation of cosmic strings or initiation of a period of exponential inflation.

A cosmic string is an object that can be obtained from an infinitely concentrated distribution of matter, with linear density of mass μ [2]. In the case of a certain distribution being located on the z-axis, the energy-momentum tensor, in cylindrical coordinates, is given by

$$T_{\nu}^{\beta} = \mu\, diag(1,0,0,1)\delta^{(2)}\left(\vec{r}\right). \tag{8}$$

Here, $\delta^{(2)}\left(\vec{r}\right)$ is a two-dimensional Dirac delta function. Geometrically, a topological defect can be characterized by a space-time whose metric associated with this defect has the corresponding Riemann-Christoffel tensor null at all points, except for the defect, i.e., the space-time has conical singularity. In other words, it may be characterized by a bending tensor, which is proportional to a delta function supported on the defect.

We want that the Eq. (8) generates a geometry with cylindrical symmetry. For that, our goal is to find a solution to Einstein's equations describing the gravitational field of an ideal cosmic string with linear mass density μ along the z-axis. In this sense, the string will have no dependence over time, so it is a temporal invariant. We will also admit a symmetry of the string in relation to the azimuth angle, and finally that it remains invariant by boosts. Thus, the most general line element, in cylindrical coordinates, which exhibits such symmetry and maintains invariance by boosts transformations along the z-axis, is given by

$$ds^2 = A^2(r)dt^2 - dr^2 - B^2(r)d\phi^2 - A^2(r)dz^2. \tag{9}$$

Using Eq. (1), taking into account the metric tensor given in Eq. (9), we can calculate the Christoffel symbols and obtain a set of non-linear differential equations given by

$$R_t^t = R_z^z = \frac{A''(r)}{A(r)} + \frac{A'(r)B'(r)}{A(r)B(r)} - \left(\frac{A'(r)}{A(r)}\right)^2, \tag{10}$$

$$R_r^r = 2\frac{A''(r)}{A(r)} + \frac{B'(r)}{B(r)}, \tag{11}$$

$$R_r^r = \frac{B''(r)}{B(r)} + 2\frac{A'(r)B'(r)}{A(r)B(r)}. \tag{12}$$

Solving these equations, we get the following solutions

$$A'(r) = \frac{d}{dr}A(r) = 0; \; \frac{BI'(r)}{B(r)} = \frac{1}{B(r)}\frac{d^2}{dr^2}B(r) = -8\pi\mu. \tag{13}$$

The above solution provides the following line element [2, 14]

$$ds^2 = dt^2 - dr^2 - (1-4\mu)d\phi^2 - dz^2. \tag{14}$$

Redefining the angular coordinate in Eq. (14), where we use the substitution $\phi' = \phi/q$ with $q^{-1} = (1-4\mu)$, we have

$$ds^2 = dt^2 - dr^2 - d\phi'^2 - dz^2, \tag{15}$$

where the angular coordinate varies in the range $\left(0, \frac{2\pi}{q}\right)$, so that space-time is now locally flat except for $r = 0$, which means except under the defect. This line element, from a global point of view, corresponds to Minkowski's space-time minus one piece subtended by the angle $8\pi\mu$. The quantity μ has great importance in string theory since it characterizes the intensity of the gravitational interaction and its value obtained from the Great Unification Theories is comprised in the order of 10^{-6} [28, 29]. Then, space-time generated by a cosmic string has the shape of a cone in the perpendicular plane to the string. Being flat itself, it satisfies Einstein's equations in every region where $T^{\beta}_{\nu} = 0$.

The effect of the string is therefore to introduce a deficit in the azimuthal angle given by $\Delta\phi = 8\pi\mu$, generating in the surface (t, z) = constants, a conical geometry instead of a flat geometry, which will be pointed in the limit of the string internal structure going to zero. In this case, the corresponding space-time is conic and best described in cylindrical coordinates due to the symmetry of the problem. The geometry described above has many interesting features, such as:

- Absence of Newtonian gravitational potential although this does not imply the absence of gravitational effects, that is, a particle placed in the presence of a cosmic string will not be attracted to it, whatever the order of magnitude of the mass density of the string, which is quite different from that predicted by Newton's gravitational string of matter; in other words, the cosmic strings have zero gravitational potential [30].

- It can act as a gravitational lens as shown in **Figure 2**, that is, due to the conic nature of space-time around the cosmic string, double images of objects located behind the string can be formed in relation to an observer [2].

- Gravitational analog of the Aharonov-Bohm effect, due to the movement of test particles in space-time of cosmic strings through the study of geodesics [31].

- Electrostatic self-interaction [13] that arises due to the gravitational field inducing a curvature in space-time, and this curvature causes distortions in the field lines of the electrostatic potential generated by a charged particle, causing this particle to undergo a finite force upon itself.

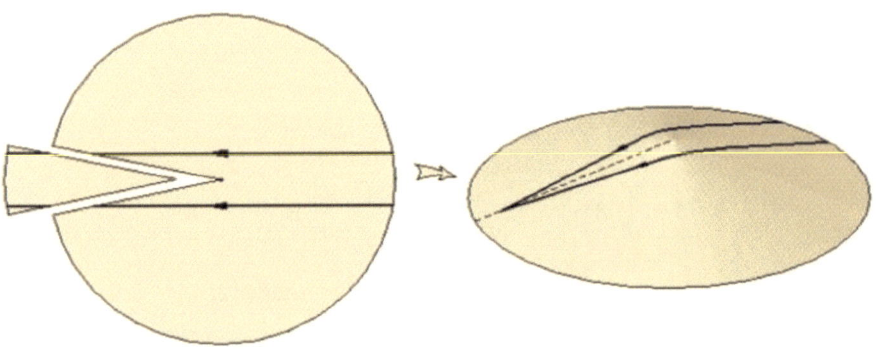

Figure 2.
Representation of the light way coming from the infinity and "curving" due to the presence of a cosmic string.

4. The Higgs mechanism

Most of the symmetries observed in nature are not exact. For example, Isospin is not an exact symmetry of nature, because the proton and the neutron do not have the same mass. One way to study symmetry breaks in field theory with symmetry breaking is to introduce the Lagrangian terms with small coefficients that explicitly perform the break. In this section, we will be interested in a symmetry breaking which the Lagrangian is symmetric under the action of a group of transformations but the state of less energy is not.

To understand how spontaneous symmetry breaking appears in many Abelian field theories, we will start considering the simple case, that is, the Lagrangian for a complex scalar field given by

$$L = \partial_\nu \varphi \partial^\nu \varphi^* - V(|\varphi|), \tag{16}$$

where $V(|\varphi|) = \mu^2 \varphi\varphi^* + \lambda(\varphi^* \varphi)^2$, being λ the self-coupling constant. To this theory, making the transformation over the scalar field is as follows

$$\varphi \to \varphi' = \varphi\, e^{iq\alpha} \; and \; \varphi^* \to \varphi'^* = \varphi^* e^{-iq\alpha}. \tag{17}$$

Here, the parameter α does not depend on the point, and we can see that the derivative in Eq. (16) goes to

$$\partial_\nu \varphi \to \partial_\nu \varphi' = (\partial_\nu \varphi)\, e^{iq\alpha} \; and \; \partial_\nu \varphi^* \to \partial_\nu \varphi'^* = (\partial_\nu \varphi^*)\, e^{-iq\alpha\wedge} = U(\partial_\nu \varphi). \tag{18}$$

Putting Eqs. (17) and (18) into Eq. (14), we see that

$$L \to L' = L. \tag{19}$$

As we may see, these transformations under the fields keep the Lagrangian unchanged. The transformations over the fields and their derivatives that do not depend on the point are named global gauge transformation.

On the other hand, let us consider that the parameter α now depends on the point, it means, $\alpha \equiv \alpha(x) \to U(x) = e^{iq\alpha(x)}$. These kinds of transformation are called local gauge transformation.

This way the transformation over the derivatives, Eq. (18), becomes

$$\partial_\nu \varphi \to \partial_\nu \varphi' = \partial_\nu \left(\varphi\, e^{iq\alpha(x)}\right) = (\partial_\nu \varphi) e^{iq\alpha(x)} + iq(\partial_\nu \alpha(x))\varphi. \tag{20}$$

As we can see, the field derivative does not transform as the field itself. The second term that appears in Eq. (20) turns the Lagrangian as not invariant by these transformations over the fields. This way, to turn this theory unchanged by transformations where the parameter now depends on the point, we have to add new fields called "compensating fields," $A_\nu(x)$. Doing this we also have to redefine the derivative concept, and this way, we have

$$\partial_\nu \to D_\nu = \partial_\nu + iqA_\nu(x). \tag{21}$$

In Eq. (21), we have the covariant derivative. Now, under transformations over the fields, the fields derivative will transform itself like the own fields, which means

$$D_\nu \varphi \to (D_\nu \varphi)' = U(x) D_\nu \varphi. \tag{22}$$

Hence, the total Lagrangian will change by the addiction of the dynamic of these "compensating fields" and its dynamic is given by the term $L(A_\nu)$ where we have only A_ν interacting among itself, this way we get

$$L = L(\varphi, \varphi^*, D_\nu\varphi, (D_\nu\varphi)^*) + L(A_\nu), \tag{23}$$

where

$$L(\varphi, \varphi^*, D_\nu\varphi, (D_\nu\varphi)^*) = D_\nu\varphi(D^\nu\varphi)^* - V(|\varphi|) \tag{24}$$

$$L(A_\nu) = -\frac{1}{4}F^{\mu\nu}F_{\mu\nu}. \tag{25}$$

Note that $F_{\mu\nu} = \partial_\mu A_\nu - \partial_\nu A_\mu$ is the Maxwell electromagnetic tensor, and the "compensating field" is the four-vector potential of the electromagnetism, and this way, the parameter q is the electron charge. In Eq. (23), we have a U(1) invariant theory that couples photons with the charged matter. This theory is the known quantum electrodynamics theory.

In general, the Higgs-Kibble mechanism is a process that generates mass for the gauge fields in this theory. Taking into account Eq. (23) with the parameters $\lambda > 0$ and $\mu^2 < 0$, this theory presents the spontaneous symmetry breaking. In this case, there exist a "ring" of degenerated vacuum states given by the minimal potential. This "ring" of degenerated vacuum values is parameterized as

$$\varphi_0 = \sqrt{\frac{|\mu^2|}{2\lambda}}e^{i\Lambda}. \tag{26}$$

The study around a vacuum value state can be done by taking the scalar field

$$\varphi = \frac{1}{\sqrt{2}}[v + \eta(x) + i\xi(x)]. \tag{27}$$

Being $v = \sqrt{|\mu^2|/\lambda}$, substituting Eq. (27) in Eq. (23), we have

$$L = -\frac{1}{4}F^{\mu\nu}F_{\mu\nu} - \frac{q^2v^2}{2}A_\nu A^\nu + \frac{1}{2}(\partial_\nu\eta)^2 + \frac{1}{2}(\partial_\nu\xi)^2 - \lambda v^2\eta^2 - qvA_\nu\xi. \tag{28}$$

The term $A_\nu A^\nu$ that appears in Eq. (28) shows that the gauge field now acquires mass. Besides that we also can see in Eq. (28) that a massive scalar field, η, with mass $m_\eta^2 = 2\lambda v^2$ and a Goldstone scalar field appear. However, the Goldstone scalar field does not present physical relevance and may be reabsorbed through a gauge field redefinition. Taking the gauge field redefinition given by

$$B_\nu = A_\nu - q\partial_\nu\xi, \tag{29}$$

we may rewrite Eq. (28) as

$$L = -\frac{1}{4}F^{\mu\nu}F_{\mu\nu} - \frac{q^2v^2}{2}B_\nu B^\nu + \frac{1}{2}(\partial_\nu\eta)^2 - \lambda v^2\eta^2, \tag{30}$$

where

$$F_{\mu\nu} = \partial_\mu B_\nu - \partial_\nu B_\mu = \partial_\mu A_\nu - \partial_\nu A_\mu. \tag{31}$$

The B_ν field presents mass $m_B = qv$, non-vanishing.

4.1 Topological defects

Topological defects are stable configurations of matter formed during phase transitions in the primordial universe. As already mentioned, during the early phases of the Universe, the material components are in physical states characterized by high degrees of symmetry and it is thought that the interactions will be unified. The cooling of the Universe, due to expansion, promotes the conditions for some of these symmetries to break, it is said, spontaneously.

This happens in much the same way as a pencil which, standing vertically and only resting on its sharp beak, drops down on a flat, oriented surface in any direction. The symmetry of rotation that exists around the axis of the pencil vanishes and, furthermore, the point where the tip was supported separates all possible positions from the topped pencil and is said to be a topological defect. (A classic example of a break in symmetry is the ferromagnetic transitions in Landau theory.) According to the types of symmetries that are broken, various types of topological defects may form, including walls, cosmic strings, monopoles, and textures. The type of defect formed is determined by the symmetry properties of the material and the nature of the phase transition.

To describe the idealized cosmic strings, i.e., static cosmic strings with infinite matter distribution along the z-axis and whose internal structure may be negligible, we will use the Nielsen and Olesen model. In this sense, by coupling the energy-momentum tensor associated with this theory to the Einstein field equations of general relativity, we study the influence of this model on space-time geometry. In fact, Laguna [10] and Garfinkle [9] did this, and in their works, they had shown that the space-time generated by the Nielsen-Olesen model was equivalent to space-time generated by a cosmic string. Thus, for a better understanding of the nature of a cosmic string, it is necessary to understand a little about models in field theory with spontaneous break of symmetry, as with the model proposed by Nielsen and Olesen.

Domain walls are two-dimensional objects that form when a discrete symmetry is broken during a phase transition. A network of walls effectively divides the Universe into several "cells." This type of defect has some very peculiar properties, one being that the gravitational field of a wall is repulsive rather than attractive. These objects may be represented as follow in **Figure 3**.

Cosmic strings are one-dimensional objects that form when an axial or cylindrical symmetry is broken. They are very thin and can extend along the visible Universe. These objects may be represented as follow in **Figure 4**.

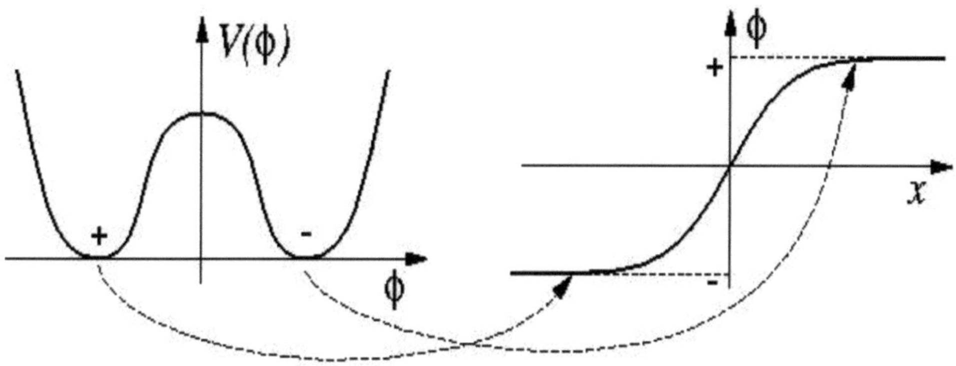

Figure 3.
Domain walls associated with models where there is more than a minimum.

Monopoles have dimension zero, that is, are punctual, and form when a spherical symmetry is broken. In field theory with non-abelian gauge symmetriy broken may appear defects like magnetic monopole. These objects may be represented as follow in **Figure 5**.

Whenever there is the possibility that cosmic strings or other topological defects form in a cosmological phase transition, they actually form. This circumstance had been first pointed out by Kibble, and therefore, in a cosmological context, the process of the formation of defects became to be known as the "Kibble mechanism" [1].

One fact regarding the universe inflation period is that the causal effects in the early universe can only propagate at the speed of light c. This means that in the instant t, regions of the Universe separated more than a distance $d = ct$ cannot know anything about each other. In a phase transition with symmetry breaking, different regions of the Universe will fall into different minimum potentials. This way, we

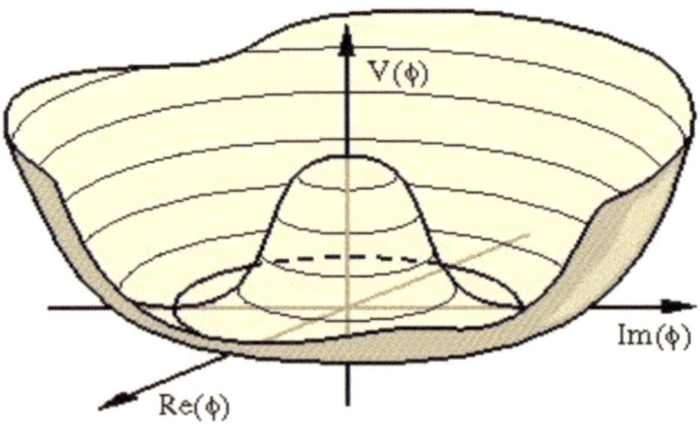

Figure 4.
Cosmic strings associated with models in which a set of minimums is not connected.

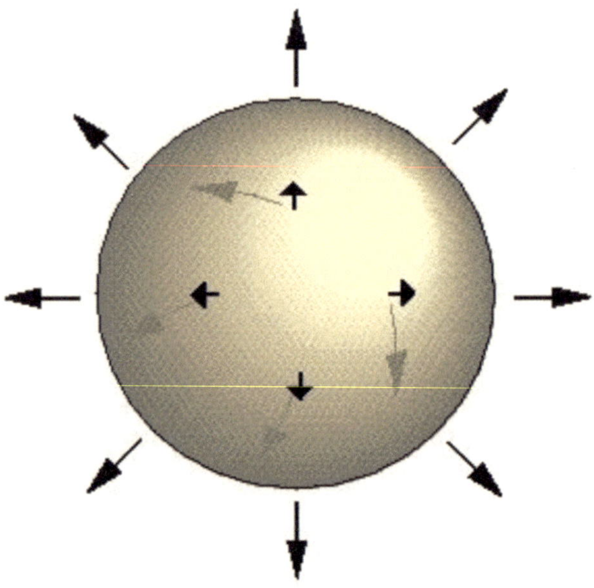

Figure 5.
Representation of a magnetic monopole defect. They are expected to be supermassive and have a magnetic charge.

actually think that topological defects are precisely the "boundaries" between these regions corresponding to different minimum potentials and their formation is thus an inevitable consequence of the phase transition.

4.2 Vortex model in field theory

The model proposed by Nielsen and Olesen for Abelian vortices, in the context of general relativity, generates a geometric structure similar to that of a cosmic string. In this sense, this object is a strong candidate to describe mathematically the cosmic strings; that is, they are strong candidates for the sources for this type of defect. However, Nielsen and Olesen, starting from a relativistic theory of fields, in 1973, have shown that it is possible to obtain solutions of vortices [8] starting from the Lagrangian density of the Abelian Higgs model, which is expressed by

$$L = -\frac{1}{4}F^{\beta\nu}F_{\beta\nu} + D_\nu\varphi(D^\nu\varphi)^* - \mu^2\varphi\varphi^* - \lambda(\varphi\varphi^*)^2. \quad (32)$$

Note that $D_\nu = \partial_\nu + ieA_\nu$ is the covariant derivative, $F_{\beta\nu} = \partial_\beta A_\nu - \partial_\nu A_\beta$ is the electromagnetism Maxwell's tensor, and $\lambda(\varphi\varphi^*)^2$ is the auto-interaction term; when this term is put, this theory starts to present a infinity degenerated vacuum, i.e., the theory has infinite states of lower energy, which satisfies the condition $|\varphi|^2 = m^2/(2\lambda)$. This way, for a particular choice vacuum configuration $\varphi = \sqrt{\frac{m^2}{2\lambda}}$, the local gauge symmetry is broken.

It is known that the action for this theory is written as

$$S = \int d^4x\, L\left(\varphi, \varphi^*, D_\nu\varphi, (D_\nu\varphi)^*, A_\mu, A_\mu^*\right). \quad (33)$$

In Eq. (17) using the Hamilton's principle, we get the following equations of motion. For $\varphi(x)$, we get

$$\frac{\partial L}{\partial\varphi} - \partial_\mu\left(\frac{\partial L}{\partial(\partial_\mu\varphi)}\right) = 0. \quad (34)$$

For A_μ, we have

$$\frac{\partial L}{\partial A^\mu} - \partial^\alpha\left(\frac{\partial L}{\partial(\partial^\alpha A^\mu)}\right) = 0. \quad (35)$$

Now using Eq. (16) into Eqs. (18) and (35), we have the following system of differential equations

$$\partial^\nu F_{\beta\nu} = j_\beta = -\frac{ie}{2}(\varphi^*\partial_\beta\varphi - \varphi\partial_\beta\varphi^*) - e^2A_\beta\varphi\varphi^* \quad (36)$$

$$D_\nu D^\nu\varphi = \lambda\varphi\left(\varphi\varphi^* - \frac{m^2}{\lambda}\right). \quad (37)$$

For a vortex in the z-direction, the components associated with the vector potential, in the Cartesian coordinate system, are $A^\mu = \left(0, A_x, A_y, 0\right)$. For this configuration, the component of the tensor $F_{\beta\nu}$ that interests us is F_{12}, because from it we can calculate the flux that passes through the plane (x, y). Parametrizing the

Higgs field by $\varphi = |\varphi| \exp{(i\chi)}$, the flux, Φ, passing through an area bounded by a closed curve C, is given by

$$\Phi = \int dxdyF_{12} = \oint_C dx^i A_i = -\frac{1}{q}\oint_C dx^i \partial_i \chi. \tag{38}$$

Here, we use, in Eq. (38), the fact that the line integral is carried out on the closed curve C, very far from the magnetic flux and that $j_\mu = 0$. The equations of motion presented in Eqs. (36) and (37) are coupled differential equations in first order that are hard to find solutions. However, the standard procedure to solve these equations, at least numerically, is to assume the following cylindrical *ansatz*, with symmetry along the z-axis for the fields [8]

$$\mathrm{A}_\mu = \left(0, 0, A(r)\hat{\theta}, 0\right) \textit{ and } \varphi(r, \theta) = f(r)e^{in\theta}. \tag{39}$$

This procedure reduces Eqs. (36) and (37) to

$$-\frac{1}{r}\frac{d}{dr}\left(r\frac{d}{dr}f(r)\right) + \left[\left(\frac{n}{r} - qA(r)\right)^2 + \lambda\left(f^2(r) - \frac{m^2}{\lambda}\right)\right] = 0, \tag{40}$$

$$-\frac{d}{dr}\left(\frac{1}{r}\frac{d}{dr}A(r)\right) + \left(q^2A(r) - \frac{nq}{r}\right)f(r) = 0. \tag{41}$$

There exist no analytical solutions to these equations. On the other hand, we can find many vortex properties by general and numerical considerations under both equations. From the general point of view, it is possible to show that these equations present solutions as asymptotically well-defined. For points closer to its nucleus, we have $f(r) \approx A(r) \to 0$. For points pretty distant from the vortex nucleus, it is observed that the functions $f(r)$ and $A(r)$ may be approximated in first order to

$$f(r) \to \frac{m}{\sqrt{\lambda}} \textit{ and } A(r) \to \frac{n}{qr}. \tag{42}$$

By using computational methods we can solve numerically Eqs. (40) and (41), and in **Figure 6**, we can see their behavior.

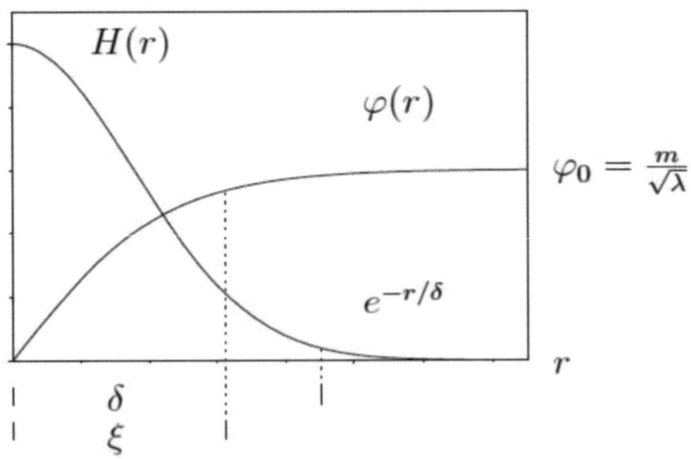

Figure 6.
$H(r)$ and $\varphi(r)$ represent, respectively, the behavior of the magnetic field and the scalar field.

From **Figure 6**, we can see that two mass scale come up, first of them is $m_s = \sqrt{2}m$ that is related with the mass of the scalar field dislocated; it means $\varphi' = \varphi - m/\sqrt{\lambda}$. The second one is related with the photon mass, $m_v = qm/\sqrt{\lambda}$, remember that the photon acquires mass because of the Higgs mechanism. Note that two length scales also appear in **Figure 6**. The first one $\delta = 1/m_v$ is related with the range of the electromagnetic field. The second latter, $\xi = 1/m_s$, is related with the space scale for the Higgs field arrive its own vacuum value.

In the literature Eqs. (23) and (24) form a system of coupled-equations and this system do not have exact solutions, but asymptotically we may solve these equations. The solutions that present finite linear density of energy, follow reference [32], are given by

$$f(r) \to \frac{m}{\sqrt{\lambda}}\left(1 - k\, e^{-\frac{r}{\xi}}\right) \; and \; A(r) \to \frac{n}{qr}\left(1 - k\, e^{-\frac{r}{\delta}}\right), \tag{43}$$

where k is a constant of proportionality.

On the other hand, Garfinkle [9], in 1985, studied the gravitational effects associated with the vortices of Nielsen and Olesen. For this purpose, he used the energy tensor, $T_{\beta\nu}$, obtained from the Lagrangian of the Abelian Higgs model, Eq. (16). In the context of the general relativity, he used this tensor as source of the Einstein equations. In this case, a static metric, with cylindrical symmetry, can be written as

$$ds^2 = e^a dt^2 - dr^2 - e^c d\phi^2 - e^b dz^2, \tag{44}$$

where a, b, and c are functions of the radio r satisfying the relations

$$a(0) = b(0) = 0 \; and \; \lim_{r\to 0} \frac{e^c}{r^2} = 1. \tag{45}$$

Given the metric, Eq. (44), solving the Einstein field equations for the energy-momentum tensor of Nielsen and Olesen, Garfinkle had found, as in flat space-time, symmetrically cylindrical static solutions which he represented as vortices. It also showed that, asymptotically, the space-time around a vortex become the Minkowski space-time minus a slice corresponding that one shown in **Figure 2**. This means that, asymptotically, the vortex can be seen as a cosmic string containing a magnetic field around it.

5. Conclusions

Throughout this work, we introduced some reasons why cosmic string-like topological defects have been studied in energy physics and condensed matter. In fact the quantum effects on the fields of matter are caused due to the non-trivial topology of these objects giving rise to polarization effects. By understanding the vacuum as a state of lower energy, the effects of vacuum polarization can be understood as changes in the scale of this lower energy. Such effects in quantum field theory are seen by calculating the vacuum expected values, VEV, of certain observables, such as the induced current density [23] and the energy-momentum tensor of the matter fields [19] induced. These observables serve as sources for the Maxwell equations in the case of induced current density and for the Einstein equations in the case of the energy-momentum tensor. In the latter case, the source of the Einstein equations no longer consists of the classical energy-momentum tensor, $T_{\mu\nu}$, but rather the energy-momentum quantizer, $\langle T_{\mu\nu} \rangle$, which will result in certain fixes in the metric tensor [33].

We have also seen that in an inertial frame, the space-time is described by the Minkowski metric, Eq. (1), which consists of a singular and diagonal metric. However, when we move to accelerated frames, the metric becomes point dependent, consisting of a set of ten space-time coordinate functions, containing all information about the geometry of the range. In this way, we can see that accelerated frames are equivalent to gravitational fields, so that gravitational effects can be described by the metric tensor, $g_{\mu\nu}(x)$. Thus, the gravitation may be understood as a deviation in the metric of the flat space-time. Moreover, this metric is not fixed arbitrarily but will depend on the distribution of local matter.

Furthermore, the cosmic string is an object whose density of matter is infinitely concentrated in a line whose mass density is μ. With this object, which can be described by the energy-momentum tensor given in Eq. (7), the deformation caused in the space-time is conical and the metric described by this density of matter is given by Eq. (15), which consists of a Minkowskian metric with cylindrical symmetry, less than a slice equal to $8\pi\mu$, which corresponds to the planar angle deficit orthogonal to the axis of symmetry of the cosmic string.

Finally, we have seen that such idealized objects can be described through the Abelian vortices models proposed by Nielsen and Olesen. They showed that by the abelian Higgs model, Eq. (32), assuming a cylindrical *ansatz*, Eq. (39), It is possible the obtaining a set of two coupled second order differential equation, as it was showed in Eqs. (40) and (41), although they do not have a closed analytic form, but that may be obtained numerical and asymptotic solutions, Eq. (27). In this way, it is observed that two length scales appear naturally from this theory. One associated with the inverse of the mass of the scalar field, $\xi \equiv 1/m_s$, and the other one related to the inverse of the mass of the vector field, $\delta \equiv 1/m_v$, which acquires mass due to the mechanism of Higgs. Also in the scope of the Abelian vortices, Linet [13] and Garfinkle [9], starting from the energy-momentum tensor associated to the Nielsen and Olesen model as the source of the Einstein field equations, they obtained a metric associated to this model, and they found a metric described by a cosmic string. The internal structure of this object is delimited by the scale of energy in which the Higgs field decays to its vacuum value.

Acknowledgements

I would like to thank Colégio de Aplicação for the technical and logistical support. I also thank the Coordenador de Educação Básica of the CAp. I thank Cassia for the emotional support that comforted me when I needed.

Conflict of interest

The authors declare no conflict of interest.

Author details

Mikael Souto Maior de Sousa[1*] and Anderson Alves de Lima[2]

1 Colégio de Aplicação, UFRR, Boa Vista, Brazil

2 Centro de Formação de Professores, UFCG, Campina Grande, Brazil

*Address all correspondence to: mikael.souto@ufrr.br

References

[1] Kibble TWB. Topology of cosmic domains and strings Journal of Physics A. 1976;**9**:1387. DOI: 10.1088/0305-4470/9/8/029/pdf

[2] Vilenkin A, Shellard EPS. Cosmic String and Other Topological Defects. Cambridge: Cambridge University Press; 1994

[3] Berezinski V, Hnatyk B, Vilenkin A. Physical Review D. 2001;**64**:043004. DOI: 10.1103/PhysRevD.64.043004

[4] Damour T, Vilenkin A. Physical Review Letters. 2000;**85**:3761. DOI: 10.1103/PhysRevLett.85.3761

[5] Bahttacharjee P, Sigl G. Physics Reports. 2000;**327**:109

[6] Schrieffer JR. Superconductivity. New York: Benjamin; 1964

[7] Abrikosov AA. Zhurnal Eksperimental'noi i Teoreticheskoi Fiziki. 1957;**32**:1442

[8] Nielsen NB, Olesen P. Nuclear Physics B. 1973;**61**:45. DOI: 10.1016/0550-3213(73)90350-7

[9] Garfinkle D. Physical Review D. 1985;**32**:1323. DOI: 10.1103/PhysRevD.32.1323

[10] Laguna-Castillo P, Matzner RA. Physical Review D. 1987;**35**:2933. DOI: 10.1103/PhysRevD.35.2933

[11] Bogomolny EB. Soviet Journal of Nuclear Physics. 1976;**24**:861

[12] Prassad MK, Sommerfield CM. Physical Review Letters. 1975;**35**:760. DOI: 10.1103/PhysRevLett.35.760

[13] Linet B. Physical Review D. 1986;**32**: 1833. DOI: 10.1103/PhysRevD.32.2511

[14] Bezerra de Mello ER, Bezerra VB, Saharian AA, Harutyunyan HH. Physical Review D. 2015;**91**:064034. DOI: 10.1103/PhysRevD.91.064034

[15] Helliwell TM, Konkowski DA. Physical Review D. 1986;**34**:1918. DOI: 10.1103/PhysRevD.34.1918

[16] Linet B. Physics Letters A. 1987;**124**: 240. DOI: 10.1016/0375-9601(87)90629-3

[17] Frolov VP, Serebriany EM. Physical Review D. 1987;**35**:3779. DOI: 10.1103/PhysRevD.35.3779

[18] Dowker JS. Physical Review D. 1987;**36**:3095. DOI: 10.1103/PhysRevD.36.3095

[19] Davies PCW, Sahni V. Classical and Quantum Gravity. 1988;**5**:1. DOI: 10.1088/0264-9381/5/1/009

[20] Dowker JS. Physical Review D. 1987;**36**:3742. DOI: 10.1103/PhysRevD.36.3742

[21] Maior de Sousa MS, Ribeiro RF, Bezerra de Mello ER. Physical Review D. 2017;**95**:045005. DOI: 10.1103/PhysRevD.95.045005

[22] Bezerra de Mello ER, Bezerra VB, Saharian AA, Tarloyan AS. Physical Review D. 2008;**78**:105007. DOI: 10.1103/PhysRevD.78.105007

[23] Maior de Sousa MS, Ribeiro RF, Bezerra de Mello ER. Physical Review D. 2016;**93**:043545. DOI: 10.1103/PhysRevD.93.043545

[24] Bezerra de Mello ER. Classical and Quantum Gravity. 1994;**11**:1415. DOI: 10.1088/0264-9381/11/6/006/meta

[25] Bezerra de Mello ER, Saharian AA. European Physical Journal C: Particles

and Fields. 2013;**73**:2532. DOI: 10.1140/epjc/s10052-013-2532-5

[26] Spinelly J, Bezerra de Mello ER. Classical and Quantum Gravity. 2003; **20**:874. DOI: 10.1088/0264-9381/20/5/307/meta

[27] Carmeli M. Classical Fields: General Relativity and Gauge Theory. New York: John Wiley and Sons; 1982

[28] Hobbs GB, Bailes M, Bhat NDR, Burke-Spolaor S, et al. Publications of the Astronomical Society of Australia. 2009;**26**:103

[29] Smith AG. In: Gibbons GW, Hawking SW, Vachaspati T, editors. Proceedings of the Cambridge Workshop. Cambridge: Cambridge University Press; 1990. p. 542

[30] Hooft G. Communications in Mathematical Physics. 1988;**117**:685. DOI: 10.1007/BF01218392

[31] Dowker JS. Nuovo Cimento. 1967; **52**:129. DOI: 10.1007%2FBF02710657

[32] Ma W, Pagels H. Physical Review C. 1978;**36**:137. DOI: 10.1016/0370-1573(78)90208-9

[33] Birrell ND, Davies PCW. Quantum Fields in Curved Space. Cambridge: Cambridge University Press; 1982

Chapter 6

Mystery of the Missing Antimatter

Golden Gadzirayi Nyambuya

"Nothing is too wonderful to be true
if it is consistent with the Laws of Nature."

— *Michael Faraday (1791-1867)*

Abstract

The Lorentz invariant Dirac equation upon which our deepest understanding of the most fundamental Quantum Mechanics is based—exhibits perfect discrete symmetries of *Charge Conjugation* (C), *Parity* (P) and *Time* (T) reversal. Not only does the Dirac equation obey these three symmetries (C, P, T), but also all the possible combinations of these symmetries, i.e., CT, CP, PT and CPT. When it comes to the C-symmetry, what this means is that—*contrary to physical and natural reality*—the Universe must contain equal portions of matter and antimatter. Obviously—this state of affairs that the Dirac equation leads to predictions that are contrary of observations—*this*—is based on the notion that the Dirac equation in its bare form as handed to us by Dirac is a correct description of reality on this front. In this chapter, we present a fundamental theoretical argument to the effect that: a symmetry violating curved spacetime version of the Dirac equation may be a perdurable solution to this long standing conundrum.

Keywords: Dirac equation: original, curved spacetime, antimatter, CP violation

1. Introduction

Perhaps persuaded and subsequently subdued by its rare, esoteric, exquisite and touchstone beauty—if one where to proceed therefrom and "religiously" believe in the verbatim predictions of the Dirac [1, 2] equation, then, they have every reason to wonder why the Universe is predominated by matter, with little—*if any*—antimatter existing naturally. The antimatter that we observe in cosmic rays is produced in the high energy collisions of matter particles. This antimatter produced in these collisions has very short life-times, lasting anything to the order of 10^{-22} s. The bare Dirac equation predicts that during the moment of creation, i.e., during the so-called *Big Bang*, matter and antimatter should have been produced in equal proportions. However, today, it vividly appears that everything we see—from the smallest life forms on Earth to the largest stellar objects in the distant heavens, is made almost entirely of matter [3]. If the assumption that matter and antimatter were created in equal portions during the so-called Big Bang, surely, something must have happened to tip the balance. As such, the answer to the question of *Where did the antimatter disappear to*? is one of the greatest challenges in physics today [3]. Physicists and scientists in general, want to know what happened to the antimatter, or why we see an acute asymmetry between matter and antimatter.

While wondering of where the antimatter disappeared to, one must on the other hand smile because if indeed this antimatter where still present, you the reader would not be able to read this book because life would not have been possible as the Universe would have been nothing more than a radiation bath, because, whenever matter and antimatter meet, they annihilate to form electromagnetic radiation. So, the Universe would have been filled with photons. Hence, it is really good that antimatter did—somehow—disappear from the Universe. Be that as it may, we still have to ask why it disappeared—why? It is in asking and seeking answers to such questions that our understanding of the inner workings of the Universe deepens.

While the term antimatter was first coined by the German-born British physicist —Sir Franz Arthur Friedrich Schuster (1851–1934) [4] on 8 August 1898 in a rather whimsical letter to the journal *Nature*; in its modern form and understanding, antimatter was first predicted in 1930 by one of the finest, greatest and towering intellectual figures in English history—Paul Adrian Maurice Dirac (1902–1984) [5]. Contrary to Sir Schuster's [4] antimatter, Dirac's [5] antimatter particles have the same mass as their matter counterparts—*albeit*—with equal but opposite electric charge. For example, an Electron's antimatter counterpart is the positively charged Positron, and that of the Proton is the negatively charged anti-Proton. Matter and antimatter particles are always produced in pairs. If they come into contact, they annihilate with one another in burst of radiation leaving behind pure energy. During the first fractions of a second of the Big Bang—physicists believe that the hot and dense Universe was buzzing with particle-antiparticle pairs *popping-in-and-out* of existence in equal proportions; thus, if matter and antimatter are created and destroyed together, it seems reasonable that the Universe should today contain nothing but a leftover relic of energy. It was the American physicist—Carl David Anderson (1905–1991) [6], who first positively detected the Position (i.e., the antimatter partner of the Electron) in the laboratory in 1932. The discovery was soon confirmed by Giuseppe Occhialini and Patrick Blackett [7], thus leading Anderson to co-receive the 1936 *Nobel Prize in Physics*.

Once it was clear that Dirac's antimatter is real, the question naturally arose— where is this antimatter? The idea that took center-stage in trying to answer this question is Sakharov's [8] hypothesis that we are going to discuss in Section 4. In a nutshell, Sakhorov's [8] hypothesis imagines a Universe perfectly symmetric in its matter-antimatter constitution and due to some subtle underlying processes, this symmetry is broken. To give a vague picture of this symmetry breaking—consider for example a coin spinning on a table. Neglecting the third side of the coin which is the edge and only considering the side—head and tail, this coin when tossed can land on either on its heads or its tails. One thing that is clear is that while the coin is still spinning before it actually lands, it cannot be defined as "heads" or "tails" until it stops spinning and falls to one side. A fair and unbiased coin has a 50–50 chance of landing on either its head or tail. Thus, if a statistically large sample of these coins is spun in exactly the same way, half should land on their heads while the other half land on tails. In pretty much the same way, half of the oscillating particles in the early Universe should have decayed as matter and the other half as antimatter.

Now—in the afore-described scenario, if say a marble rolled across a table of the spinning coins, surely, it would cause an imbalance in the 50–50 ratio of heads and tails—in simple terms, the marble would disrupt the whole system insofar as the outcome is concerned. There would an imbalance of either more heads than tails or more tails than heads. In the same fashion, some unknown mechanism [e.g., Sakhorov's [8] conditions] could have interfered with the initial balance of matter-antimatter that existed in the nascence of the Universe causing an imbalance that may have led to the majority of antimatter decaying to matter. This is the predominant thinking amongst the majority of physicists as to how the present matter-antimatter asymmetry should have come about and manifested.

The thinking that the Universe should have started-off in state of balance between matter and antimatter assumes obviously assumes that the Dirac equation in is bare form is the correct description of reality. We will argue in favor of a revision of the Dirac equation. We explore what we believe to be an equivalent of the Dirac equation on a curved spacetime. As we apply the Dirac equation in its bare form, we must be mindful that this equation does not incorporate gravity. Thus, it may very well be that, once we involve gravity, the scenario may very well be different. In actual fact, we shall do exactly that. We shall consider a set of curved spacetime Dirac equations that we proposed in the *Foundations of Physics* journal some 10 years back [9]. From these equations, we shall argue that if gravity is considered, it is possible that the Universe was created from the onset as a Universe exclusively filled with matter and with no antimatter. We think the fault squarely lies in the Dirac equation and not *Nature*—for, there is no antimatter that is missing, because it was never there to begin with.

Of the Dirac equation and its alleged fallibility, it must be said once again that this equation is one of the most successful equations in entire *History of Physics*. It is the most fundamental foundational basis of all Quantum Field Theories (QFTs) such as Quantum Electrodynamics (QED), Quantum Flavor (QFD) and Quantum Chromodynamics (QCD). It is so successful so much that, it is unimaginable for one to consider its revision—it is more like one trying to revise *Newton's Laws of Motion*. Through the centuries, *Newton's Laws of Motion* have been tested rigorously and found to be correct and the only imaginable, reasonable and logical thing one can do is to extend these into newer domains like what Einstein [10] did with the "promulgation" of his *Special Theory of Relativity* (STR).

In trying to comprehend the obvious embarrassment facing the beautiful and most successful Dirac equation—*vis*—the issue of matter-antimatter asymmetry, we must remember and be fully cognizant of the fact that the Dirac equation is an equation that is based on the flat Minkowski spacetime—gravity is not present in Minkowski spacetime hence it is not present in the Dirac equation. Thus, while this equation has been successful, this obvious fragment contradiction with physical and natural reality when it comes to the missing antimatter—this, may very well be an issue whose perdurable answer is to be found in the "complete version of the Dirac equation" and this kind of equation should most certainly be the curved spacetime version of it. In-tandem with the afore-stated philosophical approach to the "Dirac equation and the missing antimatter," the approach presented in the present chapter cannot be said to be a revision of the Dirac equation, but an extension of this equation in the domain of curved spacetime. This extended version violates C, P, T-symmetries and their combinations (i.e., CT, CP, PT and CPT), thus placing it on a sure pedestal to deliver an alternative view on this very important matter of the missing antimatter.

2. Dirac equation

In the present section, we will present an exposition of the Dirac equation and demonstrate only one of its symmetries, namely, the C-symmetry—because—this symmetry is the most relevant in our quest for the missing antimatter. As a starting point, it is important to ask "Why and how did Dirac come to discover the equation that now indelibly bares his name?" As is well known—history has recorded that Dirac embarked on the quest for the Dirac equation after it was noted that, the then and only known relativistic quantum equation—the Klein [11] and Gordon [12] Eq. (hereafter Klein-Gordon equation)—possessed both negative and positive energy solutions and in the case of the negative energy solutions, the resulting quantum probability was negative. At any rate imaginable, negative probabilities are not only meaningless, but nonsense.[1] So, Dirac embarked on his esoteric quest with in mind the aim of getting reed of the negative energy solution together with them—their negative probabilities. As fate would have it, Dirac famously achieved on tackling the latter and serendipitously failed on the former—*albeit*—it is in this failure that his agile brilliance manifested and out-shined thus placing him on a very rare and unique pedestal as one of the finest minds to ever walk the face of the planet Earth.

The Klein-Gordon equation is harnessed from the usual Einstein [10] energy-momentum equation:

$$\eta_{\mu\nu}p^{\mu}p^{\nu} = \frac{E^2}{c_0^2} - p^2 = m_0^2c_0^2, \tag{1}$$

where $\eta_{\mu\nu}$ is usual flat Minkowski metric of spacetime, p_{μ} is the four momentum, E is the energy, m_0 is the rest mass of the particle and c_0 is the speed of Light in *vacuo*. Upon a canonical quantization procedure is applied to Eq. (1), the resulting quantum mechanical equation—which is the Klein-Gordon equation—is:

$$\eta_{\mu\nu}\partial^{\mu}\partial^{\nu}\Psi = \nabla^2\Psi - \frac{1}{c_0^2}\frac{\partial^2\Psi}{\partial t^2} = \left(\frac{m_0c_0}{\hbar}\right)^2\Psi \tag{2}$$

where Ψ is the quantum mechanical wavefunction of the particle in question and ∂^{μ} is the Laplacian operator. Because the Klein-Gordon equation quadratic in the derivatives, Dirac realized that this should be the source of the negative probabilities. So, he imagined that an equation that is linear in the derivative would to be job of not only getting reed of the ugly negative probabilities, but the negative energy solutions as-well. He was right on the former and wrong on the latter.

Direct from the beautiful world of his esoteric imagination, Dirac wrote down an equation that was *in-sync* with his desideratum, i.e., an equation that is linear in the derivatives and demanded of it that when it is "squared," it would reduce to the well-known Klein-Gordon equation. Dirac's *sleight of mind* worked as he managed to obtain an equation that is so rich, it is still being studied to this day. Written in its "covariant" form, the Dirac equation is given by:

$$i\hbar\gamma_{\mu}\partial^{\mu}\psi = m_0c_0\psi, \tag{3}$$

where: $i = \sqrt{-1}$, $\hbar$ is Planck's normalized constant and:

$$\psi = \begin{pmatrix} \psi_0 \\ \psi_1 \\ \psi_2 \\ \psi_3 \end{pmatrix} = \begin{pmatrix} \psi_L \\ \psi_R \end{pmatrix}, \tag{4}$$

is the 4×1 Dirac four component wavefunction and the left- and right-handed bispinors ψ_L and ψ_R are such that:

[1] It is interesting to note that Dirac and another great mind, the American theoretical physicist—Richard Phillips Feynman (1918–1988), famously tried and monumentally failed to make sense of negative probabilities.

$$\psi_L = \begin{pmatrix} \psi_0 \\ \psi_1 \end{pmatrix} \text{and}\, \psi_R = \begin{pmatrix} \psi_2 \\ \psi_3 \end{pmatrix}, \tag{5}$$

and:

$$\gamma^0 = \begin{pmatrix} \mathcal{I}_2 & 0 \\ 0 & -\mathcal{I}_2 \end{pmatrix}, \gamma^i = \begin{pmatrix} 0 & \sigma^i \\ -\sigma^i & 0 \end{pmatrix}, \tag{6}$$

are the 4×4 Dirac gamma matrices where $\mathcal{I}_2$ and 0 are the 2×2 identity and null matrices respectively. Throughout this chapter, the Greek indices will be understood to mean: $\mu, \nu, \ldots = 0, 1, 2, 3$, while the lower case English alphabet indices: $i, j, k \ldots = 1, 2, 3$.

3. C-symmetry of the Dirac equation

For latter instructive purposes,[2] we will now demonstrate that the Dirac Eq. (3) observes C-symmetry. To that end, we shall start off in the usual manner by placing the Dirac particle ψ inside an external magnetic field whose electromagnetic four vector potential is A^{μ}_{ex}. So doing, Eq. (3) will transform to:

$$i\hbar\gamma_{\mu}\left(\partial^{\mu} + ieA^{\mu}_{ex}\right)\psi = m_0 c_0 \psi. \tag{7}$$

Now, we will have to switch the external magnetic field by reversing the electromagnetic four vector potential is A^{μ}_{ex}, i.e., $A^{\mu}_{ex} \mapsto -A^{\mu}_{ex}$. So doing, Eq. (7) will transform to:

$$i\hbar\gamma_{\mu}\left(\partial^{\mu} - ieA^{\mu}_{ex}\right)\psi = m_0 c_0 \psi. \tag{8}$$

Now, if the Dirac Eq. (3) is symmetric under electrical charge conjugation, there must exist a set of permissible mathematical operations that when applied to Eq. (8), they will lead us back to Eq. (7). The first such permissible mathematical operations is to apply the complex-operation on both-sides of Eq. (7): this complex operation will restore the sign in the coefficient of A^{μ}_{ex}, i.e.,

$$-i\hbar\gamma^{*}_{\mu}\left(\partial^{\mu} + ieA^{\mu}_{ex}\right)\psi^{*} = m_0 c_0 \psi^{*}. \tag{9}$$

Now, in-order to revert back to Eq. (7), we need to find a set of permissible mathematical operations that will remove the complex operation on γ^{*}_{μ}. This can be done because of the following *Algebra*:

$$\gamma_0\gamma_2\gamma^{*}_{\mu} = -\gamma_{\mu}\gamma_0\gamma_2. \tag{10}$$

The removal of the complex-operation on γ^{*}_{μ} is achieved by multiplying on both-sides of Eq. (9) by $\gamma_0\gamma_2$, i.e.,

$$-i\hbar\gamma_0\gamma_2\gamma^{*}_{\mu}\left(\partial^{\mu} + ieA^{\mu}_{ex}\right)\psi^{*} = m_0 c_0 \gamma_0\gamma_2\psi^{*}, \tag{11}$$

[2] That is, for the benefit of the worked presented in Section 8.

and using the fact Eq. (10), it follows that Eq. (11) will reduce to:

$$i\hbar\gamma_{\mu}(\partial^{\mu} + ieA^{\mu}_{ex})(\gamma_0\gamma_2\psi^*) = m_0c_0(\gamma_0\gamma_2\psi^*). \tag{12}$$

Now, Eq. (12) can be re-written as:

$$i\hbar\gamma_{\mu}(\partial^{\mu} + ieA^{\mu}_{ex})\psi_c = m_0c_0\psi_c, \tag{13}$$

where: $\psi_c = \gamma_0\gamma_2\psi^*$, is the antiparticle. Except for the intercharge of ψ with ψ_c, Eq. (13) is the same equation as the original Eq. (7), the meaning of which is that the Dirac equation is symmetric under charge conjugation, since the same law that applies to matter (ψ) also applies to antimatter (ψ_c).

4. Sakholov conditions

In 1967, exiled Soviet (dissident and) nuclear scientist—Andrei Dmitriev Sakharov (1921–1989), described three minimum conditions which are required for any baryogenesis to occur, regardless of the exact mechanism leading to the excess of baryonic matter. In his seminal paper which laid the foundations for all future attempts to explain the matter excess of the Universe, Sakharov [8] did not list the conditions explicitly. Instead, he described the evolution of a Universe which goes from a Baryon-excess ($\mathcal{B}$-excess) while contracting in a *Big Crunch* to an anti-*B*-excess after the resultant *Big Bang*. In summary, his three key assumptions are now known as they *Sakharov Conditions*, and these are [3]:

1. At least one *$\mathcal{B}$-number violating* process.

2. C and $\mathcal{B} = 0$-*violating* processes.

3. Interactions outside of thermal equilibrium.

These conditions must be met by any explanation in which ($\mathcal{B} = 0$) during the *Big Bang* but is very high in the present day. They are necessary but not sufficient-thus scientists seeking an explanation of the currently obtaining matter asymmetry on this basis (Sakharov conditions) must describe the specific mechanism through which baryogenesis happens. Much theoretical work in cosmology and high-energy physics revolves around finding physical processes and mechanism which fit the three Sakharov preconditions and correctly predicting the observed baryon density. Sakharov's conditions can be proven by means of Quantum Mechanics and Statistical Physics [3].

4.1 $\mathcal{B}$-number violation

We know that electric charge is a conserved quantity. The total electric charge before an interaction is always equal to the electric after the interaction. In exactly the same manner, it turns out that one can assign a charge that we call "Baryon Number" ($\mathcal{B}$-number) to quarks, and experiments have demonstrated that this charge is also conserved. All quarks carry a $\mathcal{B}$-number charge of $+1/3$, while all anti-quarks carry a $\mathcal{B}$-number charge of $-1/3$, and everything else—i.e., Leptons, intermediate gauge Bosons, etc.; carry a $\mathcal{B}$-number charge of 0. For all interaction so far observed in *Nature*, $\mathcal{B}$-number is conserved and the *World* is yet to be furnished with evidence to the contrary if indeed $\mathcal{B}$-number violation exists.

While both electric charge and $\mathcal{B}$-number are conserved physical quantities, the difference, however, between these two is that, $\mathcal{B}$-number conservation is considered an "Accidental Symmetry of Nature" because in constructing the Standard Model, one does not build in $\mathcal{B}$-number conservation explicitly. It is not a requirement for a reasonable Standard Model, but it just so happens that when one examines the evidence from numerous experiments, $\mathcal{B}$-number is always conserved somehow. Because $\mathcal{B}$-number conservation is a fact of experience, it very well may be that, there exists yet to be discovered interactions where $\mathcal{B}$-number conservation is not upheld. From an exploration stand-point, there simply is no reason why $\mathcal{B}$-number violation should be ruled out and not be considered a possibility—especially in the light of the Sakholov conditions, hence, some physicists expect that $\mathcal{B}$-number violating processes [e.g., Proton decay $p \mapsto e^{+}\pi^{0}$; $p \mapsto \mu^{+}\pi^{0}$] might exist in order to explain the matter-antimatter asymmetry.

The idea of $\mathcal{B}$-number violation is central to the so-called Grand Unified Theories, i.e., a GUT in physics is a model in particle physics in which, at exceedingly high energies $(\gtrsim 10^{16}\,\text{GeV})$, the three known gauge interactions—the Electromagnetic, Weak, and the Strong interactions, or forces which define Standard Model are merged into one single unified force or interaction where in the GUT interactions are characterized by one larger gauge symmetry group. The first true GUT was proposed by Howard Georgi and Sheldon Glashow in 1974 [13] and few months latter in the same journal by Jogesh Pati and Abdus Salam (1926–1996) [14]. There is currently no hard experimental evidence [15–18] that *Nature* is described by a GUT as Proton decay has not been observed. Without $\mathcal{B}$-number violation, it is not possible for any system to evolve from a state with: $\mathcal{B} = 0$, to a state with: $\mathcal{B} \neq 0$.

4.2 $\mathcal{P}$ and $\mathcal{CP}$-violation

The idea behind $\mathcal{CP}$-symmetry is that the equations of particle physics are invariant under mirror inversion and this leads naturally to the prediction that the mirror image of a reaction (such as a chemical reaction or radioactive decay) should occur at the same rate as the original reaction. It was not until 1956 that, along with the sacrosanct law of conservation of energy and conservation of momentum, $\mathcal{P}$-symmetry was believed to be one of the *Fundamental Geometric Conservation Laws of Nature*. After a careful and critical review of the existing experimental data by Tsung-Dao Lee and Chen-Ning Yang [19] revealed that while $\mathcal{P}$-symmetry had been verified in decays by the Strong or Electromagnetic interactions, it was untested in the Weak interaction, thus they (Lee and Yang [19]) proposed several experiments to rectify this, and simultaneously, they found a perdurable solution for this puzzle.

The first such experiment that Lee and Yang [19] proposed was the β-decay of Cobalt ^{60} nuclei whose decay is as follows:

$$^{60}_{27}\text{Co} \mapsto {}^{60}_{28}\text{Ni}^{*} + \text{e}^{-} + \overline{\nu}_{\text{e}} + 2\gamma. \tag{14}$$

Assuming parity is violated in this interaction (Eq. (14)), the Electrons ought to be ejected differently before and after the parity transformation. The landmarking experiment was soon undertaken by Wu et al. [20] the following year in 1957.

In their experiment, Wu et al. [20] aligned the spins of a sample of $^{60}_{27}\text{Co}$ with an external magnetic field. The sample was cooled to $\sim 0.003\,\text{K}$ (Houston in Lide [21] pp. 111–115) in-order to ensure that as many nuclear spins as is possible would align. Wu et al. [20] then proceeded to count the resulting decay products of the atoms along with the direction of their propagation. After a parity transformation was applied, by means of flipping the magnetic field direction, the same measurements were taken once again. The results were anything but surprising and ground-breaking. Rather than the Electrons being emitted in the same relative direction before and after the parity transformation, it was observed the Electrons had a "preferred" direction a certain direction and the effect was not small but pronounced—with a γ-ray polarization of $\sim$60%, that is to say, $\sim$60% of the γ-ray were emitted in one direction, whereas 40% were emitted in the other. If $\mathcal{P}$-symmetry was indeed conserved, Electrons would have no preferred direction of decay relative to the nuclear spin. Further, Wu et al. [20] observed that Electrons in their experiment were emitted in a direction preferentially opposite to that of the γ-rays. That is to say, most of the Electrons favored a very specific direction of decay, opposite to that of the nuclear spin. The clear meaning of these results is that $\mathcal{P}$-symmetry was violated as suggested by Lee and Yang [19], thus conclusively demonstrating that Weak interactions do indeed violate $\mathcal{P}$-symmetry.

Apart from the Co^{60} experiment, Lee and Yang [19] suggested as well that this same test of P-violation could be made in the decay of the π^+ and μ^+, i.e., the reactions:

$$\begin{aligned} \pi^+ &\mapsto \mu^+ + \nu_\mu \\ \mu^+ &\mapsto e^+ + \nu_e + \overline{\nu}_\mu. \end{aligned} \tag{15}$$

As pointed out by Garwin et al. [22], parity conservation/violation can be tested in this decay chain because if $\mathcal{P}$-symmetry is indeed broken, there should be an asymmetry in the polarization of the Muons along the direction of motion as this can be determined from the distribution of Electrons from the decay of the Muons. Experimentalists were initially skeptical that any sizeable effect of this would be measured somehow. However, after hearing of the magnitude of the asymmetry discovered by Wu et al. [20], and liaising with her in private (see e.g., Ref. [23]), Garwin et al. [22] undertook the experiment in February of 1957, the results of which they published directly after Wu et al. [20] in the same journal with the papers stuck to each other back-to-back. Ambler et al. [24], also conducted a similar experiment which coroborated Garwin et al. [22] and Wu et al.'s [20] results. In both Garwin et al. and Ambler et al. experiments, they observed angular distribution of the Electrons was as predicted by Lee and Yang [19], thus confirming that—indeed, parity is not conserved in the Weak interactions.

Having realized that $\mathcal{P}$-symmetry was indeed violated—*in an act of desperation*—it was proposed in 1957 by the great Soviet theoretical physicist—Lev Davidovich Landau (1908–1968), that $C\mathcal{P}$-symmetry was the true symmetry of *Nature* to be found between matter and antimatter, i.e., this symmetry would be conserved (see e.g., Lee et al. [25]). This proposal made use of the symmetry of a quantum mechanical system emanating from the subtle structure of Hilbert Space, that, if some symmetry say S can be found such that the combined $\mathcal{P}S$-symmetry remains unbroken, then this is the true symmetry of *Nature*. Based on this, $C\mathcal{P}$-symmetry was proposed as the desired symmetry to restore order and as we now know, this did not happen as Christenson et al. [26] demonstrated that $C\mathcal{P}$-symmetry was indeed violated. In-order to explain the matter-antimatter asymmetry, what become the central focus is the extent to which this $C\mathcal{P}$-symmetry occurred as it is believed (e.g., by Aaij et al. [27], amongst a host of many others) that this is not sufficient to account for matter-antimatter asymmetry.

4.3 Interactions outside of thermal equilibrium

The expansion [28] and ultimate accelerated expansion [29, 30] of the Universe provides the necessary platform for nonequilibrium conditions needed by the third of Sakholov's conditions requiring that there be interactions occurring outside of thermal equilibrium [3]. Actually, during most of the history of the Universe, cosmic expansion is the only source of nonequilibrium [3]. Thermal equilibrium is a time translation invariant state in which the expectation values of all observables are constant, therefore it requires a deviation from equilibrium to evolve from: $\mathcal{B} = 0$, to the desired state with: $\mathcal{B} \neq 0$ [3].

5. Current experimental and observational efforts

Gamma-ray line radiation at 511 keV is the pristine signature of Electron-Positron annihilation ($e^+e^- \mapsto \gamma\gamma$) and for the past 40 years, such radiation has been known [31] to come from the Galactic Center. Weidenspointner et al. [32] of the European Space Agency (ESA) using the INTEGRAL satellite have reported a distinct radiation asymmetry in the 511 keV line emission coming from the inner Galactic disk ($\sim 10 - 50^\circ$ from the Galactic Centre). This asymmetry Weidenspointner et al. [32] say resembles an asymmetry in the distribution of low mass X-ray binaries with strong emission at Photon energies >20 keV, indicating that they may be the dominant origin of the Positrons. This observation by Weidenspointner et al. [32] may explain the origin of a giant antimatter cloud surrounding the Galactic Center. Stellar nucleosynthesis [33–35], accreting compact objects [36–39], and even the annihilation of exotic dark-matter particles [40] have long been suggested as possible causes of this 511 keV line emission. In our view, this is interesting but does not help much in the resolution of the matter-antimatter as the question will always arise as why this antimatter is no uniformly spread.

On a recent interesting note, Neri [41] presented the first and long sought evidence for *CP*-violation in the baryon sector as this is much closer to home where it can be linked to baryon number and ultimately to $\mathcal{B}$-number violation. Neri [41] noted differences in the behavior of matter and antimatter in *K* and *B* meson decays, but not yet in any baryon decay. Such differences Neri [41] says are associated with the noninvariance of fundamental interactions under the combined *CP*-transformations, specifically *CP*-violation. In their ground breaking work, Neri [41] examined the decay products of matter and antimatter baryons (a particles containing three quarks) and looked at the spatial distribution of the resulting daughter particles within the detector. Specifically, Neri [41] looked for a very rare decay of the λ_b^0 particle (which contains an up quark, down quark and bottom quark) into a Proton and three Pions, which contain an up quark and anti-down quark, i.e., $p \mapsto \pi^-\pi^+\pi^-$ and $p \mapsto \pi^-K^+K^-$ final states. Based on data from ~ 6000 decays with a statistical significance corresponding to 3.3-sigma level including systematic uncertainties, Neri [41] find a difference in the spatial orientation of the daughter particles of the matter and antimatter λ_b^0. At a 3.3-sigma level of confidence, chances of this being a just a statistical fluctuation (and not a new property of nature) is one out of a thousand. The traditional threshold for discovery is 5-sigma level of confidence, which equates to odds of one out of more than a million.

6. Current theoretical efforts

We are of the view that theoretical effort on the problem of matter-antimatter asymmetry can be classified into two groups, the first of which are those efforts that seek to modify or extend the Dirac equation so that it is applicable on a curved spacetime and particle physics theories that try an solve this problem from with the Standard Model. We must say that—except for the efforts presented in Refs. [9, 42]

in the case of those efforts that seek to modify or extend the Dirac equation into the domain of curved spacetimes, these efforts have not been linked to the problem of matter-antimatter asymmetry.

6.1 Modified Dirac equation theories

There are several curved spacetime versions of the Dirac equation cf. [43–51] that have been proposed each with their unique taste and flavor in how it is arrived at. In our humble and modest view; save for the introduction of a seemingly mysterious four vector potential $\mathcal{A}_\mu$, what makes the curved spacetime version of the Dirac equations presented in Nyambuya [9] stand-out over other attempts is that the method used in arriving at these curved spacetime Dirac equations [9] is exactly the same as that used by Dirac [1, 2] in arriving at the Dirac equation. As will be demonstrated shortly in Section 7, this method used in Ref. [9] appears to us as the most straight forward and logical manner in which to arrive a curved spacetime version of the Dirac equation. All that has been done in Ref. [9], is to decompose the general Riemann metric $g_{\mu\nu}$ in a manner that allows us to apply Dirac's [1, 2] prescription at arriving at the Dirac equation. Apart from this attempt [9]; attempts by e.g., Refs. cf. [43–51] that have been made to date, seek a curved spacetime version of the Dirac equation not from the fundamental curved spacetime energy-momentum equation $\left(g_{\mu\nu}p^\mu p^\nu = m_0^2c^4\right)$, but take the Dirac equation as their point of departure. Our said approach (first presented in Ref. [9]) is new. In Section 7, we will present this new curved spacetime Dirac equation and proceed thereafter to demonstrate how this equation can be used to proffer a solution to this relatively long-standing matter-antimatter problem.

6.2 Modified Standard Model approach

According to Robson [52] who has developed a viable alternative to the SM-model called the Generation Model (GM) of particle physics [53–59], the matter-antimatter asymmetry problem can be solved within the framework of GM, where one can demonstrate that this asymmetry problem can be understood in terms of the composite leptons and quarks of the GM. According to Robson [52], one notes from this GM that there is essentially no matter-antimatter asymmetry in the present Universe and that the observed hydrogen-antihydrogen asymmetry may be understood in terms of statistical fluctuations associated with the complex many-body processes involved in the formation of either a hydrogen atom or an antihydrogen atom.

In Robson's [52] GM, the original antimatter created in the Big Bang is now contained within the stable composite Leptons, the Electrons and Neutrinos, and the stable composite quarks, the weak eigenstate up and down quarks that comprise the Protons and neutrons, within the hydrogen, helium and heavier atoms of the universe. Thus there is no matter-antimatter asymmetry in the present universe. However, there does exist a hydrogen-antihydrogen asymmetry where the present Universe consists predominantly of hydrogen atoms and virtually no antihydrogen atoms—and, as afore-stated—with the hydrogen-antihydrogen asymmetry understood in terms of statistical fluctuations associated with the complex many-body processes.

7. Curved spacetime Dirac equation

Our general feeling about the Dirac equation is that once it was discovered, it was taken up "very fast" and used as a most fundamental basis for building almost all—*if not all*—aspects of QFTs. In the literature, there are no real visible efforts at an attempt on a curved spacetime version of the Dirac equation in the early stages of the Dirac's discovery of the Dirac equation. The major reason why the Dirac equation was taken up fast is that, at its birth—which was also the triumphant moment of its coronation and inauguration—it astoundingly explained, in a subtly natural and exquisitely brilliant manner, the quantum mystery of the origins of spin and as-well, the then inexplicable gyromagnetic ratio of the Electron.

In the present section, we will give an exposition of the curved spacetime Dirac equation that we first presented in Ref. [9]. In this said attempt, the composite symmetric metric tensor of general relativity $g_{\mu\nu}$ consisting of 10 potentials, is decomposed into a metric that can be described by a four vector potential $\mathcal{A}_\mu$, i.e., $g_{\mu\nu} = \mathcal{A}_\mu\mathcal{A}_\nu$. Certainly, this is a reduction in complexity—i.e., from 10 potentials to only 4. At the time of our proposal, we were unsure how to interpret this description of the metric using a four vector because the only four vector potential that we know to describe a force is the Electromagnetic four vector potential.

However, if we are to still to the facts that $g_{\mu\nu}$ describes the gravitational field, then, logic dictates that the gravitational field must be capable of being described by a four vector potential as suggested by this decomposition of the metric. Latter, it become clear that indeed, the gravitational field must be describable using a four vector potential and the proof of this has been given by Hera [60] and Behera [61] in their different theorems. We will not go into these theorems but hope that the reader will visit these very important proofs and convince themselves that indeed, the gravitational field must be describable (somehow) by a four vector potential. Assuming that gravity can be described by a four vector potential, we shall now proceed to present an exposition of the aforesaid curved spacetime Dirac equation that where first presented in Ref. [9]. What is new in this exposition is that, the gravitational four vector potential $\mathcal{A}_\mu$ is consider for two cases, where in one case, these vector is a real function, while in the other, it is a complex function. It is in the case where this potential is complex that we are going to find the present proposed solution the matter-antimatter conundrum.

7.1 Real valued gravitational four vector potential

We shall begin by expanding $g_{\mu\nu}p^\mu p^\nu$ into its 16 components—*albeit*, effectively 10 components and this is because of the symmetries of the metric, i.e., $g_{\mu\nu} = g_{\nu\mu}$; the four momentum space line element equation: $g_{\mu\nu}p^\mu p^\nu = m_0^2c_0^2$, is such that:

$$\begin{aligned} g_{\mu\nu}p^\mu p^\nu = g_{00}\left(\frac{E^2}{c_0^2}\right) &+ g_{11}p^1p^1 + g_{22}p^2p^2 + g_{33}p^3p^3 \\ &+ 2g_{01}p^0p^1 + 2g_{02}p^0p^2 + 2g_{03}p^0p^3 \\ &+ 2g_{12}p^1p^2 + 2g_{13}p^1p^3 + 2g_{23}p^2p^3 = m_0^2c_0^2. \end{aligned} \tag{16}$$

The terms sandwiched between $g_{\mu\nu}p^\mu p^\nu$ and $m_0^2c_0^2$ are the expanded terms—the effective 10 components mentioned above. Upon canonical quantization, Eq. (16) becomes:

$$\begin{aligned} g_{\mu\nu}\partial^\mu\partial^\nu\Psi = g_{00}\left(\frac{\partial_0^2}{c_0^2}\right)\Psi &+ g_{11}\partial^1\partial^1\Psi + g_{22}\partial^2\partial^2\Psi + g_{33}\partial^3\partial^3\Psi \\ &+ 2g_{01}\partial^0\partial^1\Psi + 2g_{02}\partial^0\partial^2\Psi + 2g_{03}\partial^0\partial^3\Psi \\ &+ 2g_{12}\partial^1\partial^2\Psi + 2g_{13}\partial^1\partial^3\Psi + 2g_{23}\partial^2\partial^3\Psi = \left(\frac{m_0c_0}{\hbar}\right)^2\Psi, \end{aligned} \tag{17}$$

where c_0 is the speed of Light in *vacuo*. As first presented in Ref. [9]: in our quest for a curved spacetime Dirac equation which upon squaring would result in the above equation, we noted that if: $g_{\mu\nu} \propto \mathcal{A}_\mu \mathcal{A}_\nu$ (which is mathematically and logically permissible), one can write down an equivalent curved spacetime Dirac equation. We noted that there are three configurations of the metric tensor, $g_{\mu\nu}$, that would do this. So, we decided to introduce a subscript for that would identify the metric tensor with the particular configuration, where in the new metric tensor, we now have $g^{(a)}_{\mu\nu}$, with $a = (1, 2, 3)$ being the label of these three configurations. With the metric now written as $g^{(a)}_{\mu\nu}$, Eq. (17), now becomes:

$$\begin{aligned} g^{(a)}_{\mu\nu}\partial^\mu\partial^\nu\Psi = {} & g^{(a)}_{00}\left(\frac{\partial_0^2}{c_0^2}\right)\Psi + g^{(a)}_{11}\partial^1\partial^1\Psi + g^{(a)}_{22}\partial^2\partial^2\Psi + g^{(a)}_{33}\partial^3\partial^3\Psi \\ & + 2g^{(a)}_{01}\partial^0\partial^1\Psi + 2g^{(a)}_{02}\partial^0\partial^2\Psi + 2g^{(a)}_{03}\partial^0\partial^3\Psi \\ & + 2g^{(a)}_{12}\partial^1\partial^2\Psi + 2g^{(a)}_{13}\partial^1\partial^3\Psi + 2g^{(a)}_{23}\partial^2\partial^3\Psi = \left(\tfrac{m_0c_0}{\hbar}\right)^2\Psi, \end{aligned} \tag{18}$$

so that written without the expanded terms, the curved spacetime Klein-Gordon equation would be:

$$g^{(a)}_{\mu\nu}\partial^\mu\partial^\nu\Psi = \left(\frac{m_0c_0}{\hbar}\right)^2\Psi. \tag{19}$$

These three configurations [representing three configurations of spacetimes that we have called the (1) Quadratic Spacetime, (2) Parabolic Spacetime and (3) Hyperbolic Space, respectively] of the metric tensor are:

$$\left[g^{(1)}_{\mu\nu}\right] = \begin{pmatrix} +\mathcal{A}_0\mathcal{A}_0 & 0 & 0 & 0 \\ 0 & -\mathcal{A}_1\mathcal{A}_1 & 0 & 0 \\ 0 & 0 & -\mathcal{A}_2\mathcal{A}_2 & 0 \\ 0 & 0 & 0 & -\mathcal{A}_3\mathcal{A}_3 \end{pmatrix}\mathcal{I}_4, \tag{20}$$

$$\left[g^{(2)}_{\mu\nu}\right] = \begin{pmatrix} +\mathcal{A}_0\mathcal{A}_0 & +\mathcal{A}_0\mathcal{A}_1 & +\mathcal{A}_0\mathcal{A}_2 & +\mathcal{A}_0\mathcal{A}_3 \\ +\mathcal{A}_1\mathcal{A}_0 & -\mathcal{A}_1\mathcal{A}_1 & +\mathcal{A}_1\mathcal{A}_2 & +\mathcal{A}_1\mathcal{A}_2 \\ +\mathcal{A}_2\mathcal{A}_0 & +\mathcal{A}_2\mathcal{A} & -\mathcal{A}_2\mathcal{A}_2 & +\mathcal{A}_2\mathcal{A}_2 \\ +\mathcal{A}_3\mathcal{A}_0 & +\mathcal{A}_3\mathcal{A} & +\mathcal{A}_3\mathcal{A}_2 & -\mathcal{A}_3\mathcal{A}_3 \end{pmatrix}\mathcal{I}_4, \tag{21}$$

$$\left[g^{(3)}_{\mu\nu}\right] = \begin{pmatrix} +\mathcal{A}_0\mathcal{A}_0 & -\mathcal{A}_0\mathcal{A}_1 & -\mathcal{A}_0\mathcal{A}_2 & -\mathcal{A}_0\mathcal{A}_3 \\ -\mathcal{A}_1\mathcal{A}_0 & -\mathcal{A}_1\mathcal{A}_1 & +\mathcal{A}_1\mathcal{A}_2 & -\mathcal{A}_1\mathcal{A}_3 \\ -\mathcal{A}_2\mathcal{A}_0 & -\mathcal{A}_2\mathcal{A}_1 & -\mathcal{A}_2\mathcal{A}_2 & -\mathcal{A}_2\mathcal{A}_3 \\ -\mathcal{A}_3\mathcal{A}_0 & -\mathcal{A}_3\mathcal{A}_1 & +\mathcal{A}_3\mathcal{A}_2 & -\mathcal{A}_3\mathcal{A}_3 \end{pmatrix}\mathcal{I}_4. \tag{22}$$

In a general, written in a condensed form, these three metric tensors $g^{(1)}_{\mu\nu}$, $g^{(2)}_{\mu\nu}$ and $g^{(3)}_{\mu\nu}$, are such that:

$$\left[g^{(a)}_{\mu\nu}\right] = \begin{pmatrix} +\mathcal{A}_0\mathcal{A}_0 & \lambda_a\mathcal{A}_0\mathcal{A}_1 & \lambda_a\mathcal{A}_0\mathcal{A}_2 & \lambda_a\mathcal{A}_0\mathcal{A}_3 \\ \lambda_a\mathcal{A}_1\mathcal{A}_0 & -\mathcal{A}_1\mathcal{A}_1 & \lambda_a\mathcal{A}_1\mathcal{A}_2 & \lambda_a\mathcal{A}_1\mathcal{A}_3 \\ \lambda_a\mathcal{A}_2\mathcal{A}_0 & \lambda_a\mathcal{A}_2\mathcal{A}_1 & -\mathcal{A}_2\mathcal{A}_2 & \lambda_a\mathcal{A}_2\mathcal{A}_3 \\ \lambda_a\mathcal{A}_3\mathcal{A}_0 & \lambda_a\mathcal{A}_3\mathcal{A}_1 & \lambda_a\mathcal{A}_3\mathcal{A}_2 & -\mathcal{A}_3\mathcal{A}_3 \end{pmatrix}\mathcal{I}_4, \tag{23}$$

where—further: in a much more compact form, this metric can be written as:

$$g_{\mu\nu}^{(a)} = \frac{1}{2}\left(\mathcal{A}_{\mu}^{(a)}\mathcal{A}_{\nu}^{(a)} + \mathcal{A}_{\nu}^{(a)}\mathcal{A}_{\mu}^{(a)}\right) = \frac{1}{2}\left\{\mathcal{A}_{\mu}^{(a)}, \mathcal{A}_{\nu}^{(a)}\right\} \in \mathbb{R}, \tag{24}$$

where:

$$\mathcal{A}_{\mu}^{(a)} = \gamma_{\mu}^{(a)}\mathcal{A}_{\mu}, \tag{25}$$

and the 4×4 matrices $\gamma_{\mu}^{(a)}$ are such that:

$$\begin{aligned} \gamma_{0}^{(a)} &= \begin{pmatrix} \mathcal{I}_2 & 0 \\ 0 & -\mathcal{I}_2 \end{pmatrix} = \gamma^0, \\ \gamma_{k}^{(a)} &= \frac{1}{2}\begin{pmatrix} 2\lambda_a \mathcal{I}_2 & i^{\lambda_a|\lambda_a|}\sqrt{2^{|\lambda_a|}}\sigma_k \\ i^{\lambda_a|\lambda_a|}\sqrt{2^{|\lambda_a|}}\sigma_k & -2\lambda_a\mathcal{I}_2 \end{pmatrix}. \end{aligned} \tag{26}$$

The λ_a's in Eq. (26) are defined such that when:

$$a = \begin{cases} 1, & \text{then} \quad (\lambda_1 = 0): \quad \text{Quadratic Spacetime (QST)}. \\ 2, & \text{then} \quad (\lambda_2 = +1): \quad \text{Parabolic Spacetime (PST)}. \\ 3, & \text{then} \quad (\lambda_3 = -1): \quad \text{Hyperbolic Spacetime (HST)}. \end{cases} \tag{27}$$

The index "a" is not an active index as are the Greek indices. This index labels a particular curvature of spacetime, i.e., whether spacetime is flat,[3] positive or negatively curved as defined by the resulting metric $g_{\mu\nu}^{(a)}$ which is given in Eq. (23). So, in the end, the resulting and desired curved spacetime Dirac equation is:

$$i\hbar\mathcal{A}_{\mu}^{(a)}\partial^{\mu}\psi = m_0 c_0 \psi, \tag{28}$$

where it is understood that $\mathcal{A}_{\mu}$ is a real valued gravitational four vector function. In the subsequent section, we will show that the above equation with a complex valued gravitational four vector function violates C-symmetry. A violation of C-symmetry is all one needs to explain the whereabouts of the missing antimatter. Multiplication by $\left(i\hbar\mathcal{A}_{\nu}^{(a)}\partial^{\nu}\right)$ from the left on the left handside of Eq. (28) and on the right by m_0c_0 and this is on the understanding that these operators are identical —i.e., $i\hbar\mathcal{A}_{\nu}^{(a)}\partial^{\nu} \equiv m_0c_0\mathcal{I}_4$: one will lead to the curved spacetime Klein-Gordon equation provided:

$$\mathcal{A}_{\nu}^{(a)}\partial^{\nu}\mathcal{A}_{\mu}^{(a)} = 0. \tag{29}$$

Therefore, the above equation enters into this theory as a gauge condition to be met by the real-valued gravitational four-vector potential $\mathcal{A}_{\mu}^{(a)}$.

[3] By *flat*, it here is not meant that the spacetime is Minkowski *flat*, but that the metric has no off diagonal terms. On the same footing, by *positively curved spacetime*, it meant that metric has positive off diagonal terms and likewise, a *negatively curved spacetime*, it meant that metric has negative off diagonal terms.

7.2 Complex valued gravitational four vector potential

We shall now assume that the gravitational four vector field is complex valued and we shall denote it as: $\tilde{\mathcal{A}}_{\nu}^{(a)}$ and $\tilde{\mathcal{A}}_{\nu}$. In this event where we have a complex valued gravitational four vector field, the metric will have to be defined as:

$$g_{\mu\nu}^{(a)} = \frac{1}{2}\left(\tilde{\mathcal{A}}_{\mu}^{(a)\dagger}\tilde{\mathcal{A}}_{\nu}^{(a)} + \tilde{\mathcal{A}}_{\nu}^{(a)\dagger}\tilde{\mathcal{A}}_{\mu}^{(a)}\right) = \frac{1}{2}\left\{\tilde{\mathcal{A}}_{\mu}^{(a)\dagger}, \tilde{\mathcal{A}}_{\nu}^{(a)}\right\}, \tag{30}$$

where—as before:

$$\tilde{\mathcal{A}}_{\mu}^{(a)} = \tilde{\gamma}_{\mu}^{(a)}\mathcal{A}_{\mu}, \tag{31}$$

and the new 4×4 $\tilde{\gamma}$-matrices are such that:

$$\tilde{\gamma}_{0}^{(a)} = \begin{pmatrix} 0 & \mathcal{I}_2 \\ -\mathcal{I}_2 & 0 \end{pmatrix}, \qquad \tilde{\gamma}_{k}^{(a)} = \frac{1}{2}\begin{pmatrix} i^{\lambda_a|\lambda_a|}\sqrt{2^{|\lambda_a|}}\sigma_k & 2\lambda_a\mathcal{I}_2 \\ -2\lambda_a\mathcal{I}_2 & i^{\lambda_a|\lambda_a|}\sqrt{2^{|a|}}\sigma_k \end{pmatrix}. \tag{32}$$

Written in full as is the case in Eq. (23), the metric $g_{\mu\nu}^{(a)}$ for a complex valued gravitational four vector potential $\mathcal{A}_{\mu}$ is such that:

$$\left[g_{\mu\nu}^{(a)}\right] = \begin{pmatrix} +\mathcal{A}_0^{\dagger}\mathcal{A}_0 & \lambda_a\mathcal{A}_0^{\dagger}\mathcal{A}_1 & \lambda_a\mathcal{A}_0^{\dagger}\mathcal{A}_2 & \lambda_a\mathcal{A}_0^{\dagger}\mathcal{A}_3 \\ \lambda_a\mathcal{A}_1^{\dagger}\mathcal{A}_0 & -\mathcal{A}_1^{\dagger}\mathcal{A}_1 & \lambda_a\mathcal{A}_1^{\dagger}\mathcal{A}_2 & \lambda_a\mathcal{A}_1^{\dagger}\mathcal{A}_3 \\ \lambda_a\mathcal{A}_2^{\dagger}\mathcal{A}_0 & \lambda_a\mathcal{A}_2^{\dagger}\mathcal{A}_1 & -\mathcal{A}_2^{\dagger}\mathcal{A}_2 & \lambda_a\mathcal{A}_2^{\dagger}\mathcal{A}_3 \\ \lambda_a\mathcal{A}_3^{\dagger}\mathcal{A}_0 & \lambda_a\mathcal{A}_3^{\dagger}\mathcal{A}_1 & \lambda_a\mathcal{A}_3^{\dagger}\mathcal{A}_2 & -\mathcal{A}_3^{\dagger}\mathcal{A}_3 \end{pmatrix}\mathcal{I}_4. \tag{33}$$

With the $\tilde{\gamma}$-matrices defined; in-order for $g_{\mu\nu}^{(a)} \in \mathbb{R}$, the gravitational four vector potential will have to be defined in the *de* Broglie-Bohm [62–64] polar form as follows:

$$\mathcal{A}_{\mu} = \phi_{\mu}\exp\left(\frac{i\mathcal{S}}{\hbar}\right), \tag{34}$$

where: $\phi_{\mu} = \phi_{\mu}(\boldsymbol{r}, t) \in \mathbb{R}$, is a differentiable, uniform continuous, and integrable four-vector-valued function; $\mathcal{S} = \mathcal{S}(\boldsymbol{r}, t) \in \mathbb{R}$, is a zero rank scalar that is also a differential, uniform continuous and integrable function. With $\mathcal{A}_{\mu}$ defined as it defined above, it follows that:

$$\tilde{\mathcal{A}}_{\mu}^{*}\tilde{\mathcal{A}}_{\nu} = \tilde{\mathcal{A}}_{\nu}^{*}\tilde{\mathcal{A}}_{\mu} = \phi_{\mu}\phi_{\nu} \in \mathbb{R}, \tag{35}$$

hence:

$$g_{\mu\nu}^{(a)} = \frac{1}{2}\left\{\tilde{\mathcal{A}}_{\mu}^{(a)\dagger}, \tilde{\mathcal{A}}_{\nu}^{(a)}\right\} = \frac{1}{2}\left\{\tilde{\gamma}_{\mu}^{(a)\dagger}, \tilde{\gamma}_{\nu}^{(a)}\right\}\tilde{\mathcal{A}}_{\mu}^{*}\tilde{\mathcal{A}}_{\nu} = \frac{1}{2}\left\{\tilde{\gamma}_{\mu}^{(a)\dagger}, \tilde{\gamma}_{\nu}^{(a)}\right\}\phi_{\mu}\phi_{\nu} \in \mathbb{R}. \tag{36}$$

So, in the end, the resulting and desired curved spacetime Dirac equation is:

$$i\hbar\tilde{\mathcal{A}}_{\mu}^{(a)}\partial^{\mu}\psi = m_0c_0\psi, \tag{37}$$

where it is understood that $\tilde{\mathcal{A}}_{\mu}$ is to be a complex valued gravitational four vector function. In the subsequent section, we will show that the above equation with a complex valued gravitational four vector function violates C-symmetry. A violation of C-symmetry is all one needs to explain the whereabouts of the missing antimatter. Multiplication by $\left(i\hbar\mathcal{A}_{\nu}^{(a)}\partial^{\nu}\right)^{\dagger}$ from the left on the left handside of Eq. (37) and on the right by $(m_0c_0)^{\dagger}$ and this is on the understanding that these operators are identical—i.e., $\left(i\hbar\mathcal{A}_{\nu}^{(a)}\partial^{\nu}\right)^{\dagger} \equiv (m_0c_0)^{\dagger}\mathcal{I}_4$: one will be led to the curved spacetime Klein-Gordon equation provided:

$$\tilde{\mathcal{A}}_{\nu}^{(a)\dagger}\partial^{\nu}\tilde{\mathcal{A}}_{\mu}^{(a)} = 0. \tag{38}$$

Therefore—as before, i.e., as in Eq. (29), the above Eq. (38) enters into this theory as a *gauge condition* to be met by the gravitational four vector potential $\tilde{\mathcal{A}}_{\mu}^{(a)}$.

8. Symmetries of the curved spacetime Dirac equation

We will demonstrate that the complex valued gravitational four vector potential curved spacetime Dirac Eq. (37) violates C-symmetry. To that end, we shall start off in the usual manner by placing the curved spacetime Dirac particle ψ inside an external magnetic field whose electromagnetic four vector potential is A^{μ}_{ex}. So doing, Eq. (37) will transform to:

$$i\hbar\tilde{\mathcal{A}}_{\mu}^{(a)}(\partial^{\mu} + ieA^{\mu}_{ex})\psi = m_0c_0\psi. \tag{39}$$

Now, we will have to switch the external magnetic field by reversing the electromagnetic four vector potential is A^{μ}_{ex}, i.e., $A^{\mu}_{ex} \mapsto -A^{\mu}_{ex}$. So doing, Eq. (39) will transform to:

$$i\hbar\tilde{\mathcal{A}}_{\mu}^{(a)}(\partial^{\mu} - ieA^{\mu}_{ex})\psi = m_0c_0\psi. \tag{40}$$

Now, if Eq. (37) is symmetric under electrical charge conjugation, there must exist a set of permissible mathematical operations that when applied to Eq. (40), they will lead us back to Eq. (39). The first such permissible mathematical operations is to apply the complex-operation on both-sides of Eq. (39): this complex operation will restore the sign in the coefficient of A^{μ}_{ex}, i.e.,

$$-i\hbar\tilde{\mathcal{A}}_{\mu}^{*}\tilde{\gamma}_{\mu}^{(a)*}(\partial^{\mu} + ieA^{\mu}_{ex})\psi^{*} = m_0c_0\psi^{*} \tag{41}$$

Now, in-order to revert back to Eq. (39), we need to find a set of permissible mathematical operations that will remove the complex operation on the terms: $\tilde{\mathcal{A}}_{\mu}^{*}$ and $\tilde{\gamma}_{\mu}^{(a)*}$. We can remove the complex-operation on $\tilde{\gamma}_{\mu}^{(a)*}$ because of the following *Algebra*:

$$\gamma_0\gamma_2\tilde{\gamma}_{\mu}^{(a)*} = -\tilde{\gamma}_{\mu}^{(a)}\gamma_0\gamma_2. \tag{42}$$

The removal of the complex-operation on $\tilde{\gamma}_{\mu}^{(a)*}$ is achieved by multiplying on both-sides of Eq. (41) by $\gamma_0\gamma_2$, i.e.,

$$-i\hbar\tilde{\mathcal{A}}_{\mu}^{*}\gamma_0\gamma_2\tilde{\gamma}_{\mu}^{(a)*}(\partial^{\mu} + ieA^{\mu}_{ex})\psi^{*} = m_0c_0\gamma_0\gamma_2\psi^{*}, \tag{43}$$

and using the fact Eq. (42), it follows that Eq. (43) will reduce to:

$$i\hbar\tilde{\mathcal{A}}_\mu^* \tilde{\gamma}_\mu^{(a)}(\partial^\mu + ieA_{ex}^\mu)(\gamma_0\gamma_2\psi^*) = m_0c_0(\gamma_0\gamma_2\psi^*). \tag{44}$$

Now, Eq. (44) can be re-written as:

$$i\hbar\tilde{\mathcal{A}}_\mu^* \tilde{\gamma}_\mu^{(a)}(\partial^\mu + ieA_{ex}^\mu)\psi_c = m_0c_0\psi_c, \tag{45}$$

where: $\psi_c = \gamma_0\gamma_2\psi^*$, is the antiparticle. Now, in-order for the above equation to revert back to the original Eq. (39), there is need for the gravitational four vector, $\tilde{\mathcal{A}}_\mu$, to be real, i.e., $\tilde{\mathcal{A}}_\mu^* = \tilde{\mathcal{A}}_\mu$. If this condition $\left(\tilde{\mathcal{A}}_\mu^* = \tilde{\mathcal{A}}_\mu\right)$ cannot be met because A_μ is a complex valued function, then, the curved spacetime Dirac equation is not symmetric under charge conjugation, hence it will violate C-symmetry.

9. General discussion

Tremendous effort and thrust has been put on experimental and observational attempts whose aim is to procure the necessary evidence to support Sakholov's [8] hypothesis of the *sine-quo-non* conditions needed to be met in-order to explain the clearly obvious matter dominance observed in the Universe. Little or no effort has been put—let alone suggested, that, perhaps, the fault (solution) may lay in the very *Physical Law* that we have used to probe and understand the Universe and this law is the all-symmetric Dirac equation. It is quite understandable why this may be the case—the Dirac equation is so successful so much that, it is easy to be "blinded" by this success to an extent that one cannot—*with suspicion*—point the "little prickling finger" at it. In the present chapter, we have had to gather the necessary courage and temerity to do just that.

From what has been presented above, it is clear that from a theoretical standpoint, all one would need in-order to explain the missing antimatter is to proceed and henceforth make the hypothesis that the gravitational four vector, $\mathcal{A}_\mu$, is a complex field. This would mean that during the moment of creation, either matter is produced, with no antimatter, or, antimatter is produced, with no matter! Thus, in the framework of the foregoing curved spacetime version of the Dirac equation, the Universe is pristinely asymmetric in its matter-antimatter constitution right from the moment of creation. There would be no need to have the Sakholov conditions, or, any other exogenous mechanism or condition in-order for one to explain the matter-antimatter asymmetry. This alternative way at looking at this long standing problem appears to be the simplest way out of this ponderous and vexing conundrum of the missing antimatter. To accept this solution requires one to accept the proposed curved spacetime Dirac equations.

Sakholov's [8] hypothesis starts off by accepting the Dirac Eq. (3) in its bare form, the meaning of which is that it assumes a perfectly symmetric Universe which then proceeds to become asymmetric once the Sakholov's [8] conditions are met. Sakholov's [8] conditions require C, $\mathcal{CP}$, $\mathcal{B}$-number violating processes and the existence of nonthermodynamic equilibrium. The point here is that—what is needed is that a section of the symmetric matter-antimatter soup meets Sakholov's [8] criterion of having these processes and once this is the case, the Universe can then proceed from a state with: $\mathcal{B} = 0$, to a state with: $\mathcal{B} \neq 0$. In a perfectly matter-antimatter symmetric Universe, these processes may require certain physical conditions of energy and temperature in-order to trigger them, thus leading to a matter-antimatter asymmetric Universe.

However, in the suggestion being made here-in, we envisage a Universe where matter is created *via* the C-symmetry violating curved spacetime Dirac Eq. (37) where the Universe is created containing only matter and no antimatter and this will come about because of the phase factor in the gravitational four vector field: $\mathcal{A}_\mu^{(a)} = \phi_\mu e^{i\mathcal{S}/\hbar}$; that is to say, for so long as at the moment of creation, this phase factor is not equal to zero $\mathcal{S} \neq 0$, the Universe will be completely asymmetric in its matter-antimatter constitution. Even if the Universe where evolve to from a state with: $\mathcal{S} \neq 0$, to a state with: $\mathcal{S} = 0$, at a latter time in its evolution, the Universe will—*throughout its entire evolution*—still be asymmetric in its matter-antimatter constitution. In the end, no experiments will be need to find these C, $C\mathcal{P}$, $\mathcal{B}$-number violating processes. All we would need is to test the curved spacetime Dirac equation where our matter-antimatter asymmetry is being championed.

In-closing: insofar as accepting the proposed curved spacetime Dirac Eq. (37), it is important to note that the way these equations have been "derived" is exactly the same-way Dirac arrived at his equation. All we have done in this proposed curved spacetime Dirac Eq. (37) is to note that the metric tensor of spacetime $g_{\mu\nu}$, can be decomposed in such a manner that at its most fundamental and simplest level, it can be represented by a four vector $\mathcal{A}_\mu$. This gravitational four vector potential, $\mathcal{A}_\mu$, will have to represent the gravitational field. On this, one may object because the GTR—which is not only the current best model of gravitation, but the most successful model of gravitation; describes gravity as tensor field that is represented by not four, but 10 potentials.

Author details

Golden Gadzirayi Nyambuya
Department of Applied Physics, Fundamental Theoretical and Astrophysics Group, Faculty of Applied Sciences, National University of Science and Technology, Bulawayo, Republic of Zimbabwe

*Address all correspondence to: physicist.ggn@gmail.com

References

[1] Dirac PAM. The quantum theory of the Electron. Proceedings. Royal Society of London. 1928;**A117**:610-612. DOI: 10.1098/rspa.1928.0023

[2] Dirac PAM. The quantum theory of the Electron II. Proceedings. Royal Society of London. 1928;**A118**:351-361. DOI: 10.1098/rspa.1928.0056

[3] Canetti L, Drewes M, Shaposhnikov M. Matter and antimatter in the universe. New Journal of Physics. 2012; **14**(9):095012. DOI: 10.1088/1367-2630/14/9/095012

[4] A. Schuster, Potential matter: A holiday dream, Nature 58 (1898) 367. doi:10.1038/058367a0. Available from: http://adsabs.harvard.edu/abs/1898Natur..58..367S

[5] Dirac PAM. A theory of the electrons and protons. Proceedings. Royal Society of London. 1930;**A126**(801):360-365. DOI: 10.1098/rspa.1930.0013

[6] Anderson CD. The positive Electron. Physics Review. 1933;**43**:491-494. DOI: 10.1103/PhysRev.43.491

[7] Fitch VL. Elementary particle physics: The origins. Reviews of Modern Physics. 1999;**71**(2):S25-S32. DOI: 10.1103/revmodphys.71.s25

[8] Sakhorov AD. Violation of CP symmetry, C-asymmetry and baryon asymmetry of the universe. Journal of Experimental and Theoretical Physics Letters. 1967;**5**:24-27

[9] Nyambuya GG. New curved spacetime Dirac equations. Foundations of Physics. 2008;**37**(7):665-677. DOI: 10.1007/s10701-008-9226-0

[10] Einstein A. Zur Elektrodynamik bewegter Körper (On the electrodynamics of moving bodies). Annalen der Physik. 1905;**17**:891. Available from: http://www.fourmilab.ch/etexts/einstein/specrel/www/

[11] Klein O. Quantentheorie und fünfdimensionale Relativitätstheorie. Zeitschrift für Physik. 1926;**37**(12): 895-906. DOI: 10.1007/BF01397481

[12] Gordon W. Der Comptoneffekt nach der Schrödingerschen Theorie. Zeitschrift für Physik. 1926;**40**(1): 117-133. DOI: 10.1007/BF01390840

[13] Georgi H, Glashow SL. Unity of all elementary-particle forces. Physical Review Letters. 1974;**32**:438-441. DOI: 10.1103/PhysRevLett.32.438. Available from: http://adsabs.harvard.edu/abs/1974PhRvL..32..438G

[14] Pati JC, Salam A. Lepton number as the fourth "color". Physical Review. 1974;**D.10**:275-289. DOI: 10.1103/PhysRevD.10.275. Available from: http://adsabs.harvard.edu/abs/1974PhRvD..10..275P

[15] Bajc B, Hisano J, Kuwahara T, Omura Y. Threshold corrections to dimension-six proton decay operators in non-minimal SUSY SU (5) GUTs. Nuclear Physics B. 2016;**910**:1-22. DOI: 10.1016/j.nuclphysb.2016.06.017

[16] Babu KS, Kearns E, Al-Binni U, Banerjee S, et al. Baryon Number Violation. arXiv e-prints. 2013;**1**:1-39. arXiv:1311.5285

[17] Nishino H, Clark S. Search for proton decay via $p \to e^+\pi^0$ and $p \to \mu^+\pi^0$ in a large water Cherenkov detector. Physical Review Letters. 2009; **102**:141801. DOI: 10.1103/PhysRevLett.102.141801

[18] Sreekantan BV. Searches for proton decay and super heavy magnetic

monopoles. Journal of Astrophysics and Astronomy. 1984;**5**(3):251-271. DOI: 10.1007/bf02714542

[19] Lee TD, Yang CN. Question of parity conservation in weak interactions. Physical Review. 1956; **104**(1):254-258. DOI: 10.1103/physrev.104.254

[20] Wu CS, Ambler E, Hayward RW, Hoppes DD, Hudson RP. Experimental test of parity conservation in beta decay. Physical Review. 1957; **105**(4):1413-1415. DOI: 10.1103/physrev.105.1413

[21] Lide DR. A Century of Excellence in Measurements, Standards, and Technology. CRC Press; 2001

[22] Garwin RL, Lederman LM, Weinrich M. Observations of the failure of conservation of parity and charge conjugation in meson decays: The magnetic moment of the free muon. Physical Review. 1957;**105**(4):1415-1417. DOI: 10.1103/physrev.105.1415

[23] Lindley D. Landmarks: Breaking the mirror. Focus. 22. DOI: 10.1103/physrevfocus.22.19

[24] Ambler E, Hayward RW, Hoppes DD, Hudson RP, Wu CS. Further experiments on β-decay of polarized nuclei. Physical Review. 1957;**106**(6): 1361-1363. DOI: 10.1103/physrev.106.1361

[25] Lee TD, Oehme R, Yang CN. Remarks on possible noninvariance under time reversal and charge conjugation. Physical Review. 1957; **106**(2):340-345. DOI: 10.1103/physrev.106.340

[26] Christenson JH, Cronin JW, Fitch VV, Turlay R. Evidence for the 2πdecay of the K20meson. Physical Review Letters. 1964;**13**(4):138-140. DOI: 10.1103/physrevlett.13.138

[27] Aaij R et al. The LHCb collaboration, search for Baryon-number violating oscillations. Physical Review Letters. 2017;**119**:181807. DOI: 10.1103/PhysRevLett.119.181807

[28] Hubble EP. A relation between distance and radial velocity among extra-galactic nebulae. 1929;**15**:168-173. DOI: 10.1073/pnas.15.3.168

[29] Perlmutter S, Aldering G, Goldhaber G, Knop RA, Nugent P, Castro PG, et al. Measurements of Ω and Λ from 42 High-Redshift Supernovae. 1999;**517**:565-586. DOI: 10.1086/307221

[30] Riess AG, Filippenko AV, Challis P, Clocchiatti A, Diercks A, Garnavich PM, et al. Observational evidence from supernovae for an accelerating universe and a cosmological constant. 1998;**116**: 1009-1038. DOI: 10.1086/300499

[31] Leventhal M, MacCallum CJ, Stang PD. Detection of 511 keV positron annihilation radiation from the galactic center direction. The Astrophysical Journal. 1978;**225**:L11. DOI: 10.1086/182782

[32] Weidenspointner G, Skinner G, Jean P, Knödlseder J, von Ballmoos P, Bignami G, et al. An asymmetric distribution of positrons in the galactic disk revealed by γ-rays. Nature. 2008; **451**(7175):159-162. DOI: 10.1038/nature06490

[33] Fajans K, Morris DFC. Discovery and naming of the isotopes of element 91. Nature. 1973;**244**(5412):137-138. DOI: 10.1038/244137a0

[34] Clayton DD, Hoyle F. Gamma-ray lines from novae. Astrophysical Journal. 1974;**187**:L101. DOI: 10.1086/181406

[35] Prantzos N, Casse M. On the production of Al-26 by wolf-Rayet stars–galactic yield and gamma-ray line

emissivity. Astrophysical Journal. 1986; **307**:324. DOI: 10.1086/164419

[36] Ramaty R, Lingenfelter RE. Ray line astronomy. Nature. 1979;**278**(5700): 127-132. DOI: 10.1038/278127a0

[37] Guessoum N, Jean P, Prantzos N. Microquasars as sources of positron annihilation radiation. Astronomy and Astrophysics. 2006;**457**(3):753-762. DOI: 10.1051/0004-6361:20065240

[38] Totani T. A RIAF interpretation for the past higher activity of the galactic center black hole and the 511 keV annihilation emission. Publications of the Astronomical Society of Japan. 2006;**58**(6):965-977. DOI: 10.1093/pasj/58.6.965

[39] Cheng KS, Chernyshov DO, Dogiel VA. Annihilation emission from the galactic black hole. Astrophysical Journal. 2006;**645**(2):1138-1151. DOI: 10.1086/504583

[40] Böhm C, Hooper D, Silk J, Casse M, Paul J. MeV dark matter: Has it been detected? Physical Review Letters;**92** (10). DOI: 10.1103/physrevlett.92.101301

[41] Neri N. The LHCb collaboration, measurement of matter-antimatter differences in beauty baryon decays. Nature Physics. 2017;**13**(4):391-396. DOI: 10.1038/nphys4021.

[42] Nyambuya GG. On the preponderance of matter over antimatter. Journal of Modern Physics. 2014;**06**(11):1441-1451. DOI: 10.4236/jmp.2015.611148

[43] Alhaidari AD, Jellal A. Dirac and Klein-Gordon Equations in Curved Space. arXiv:1106.2236v3; 2014. pp. 1-8

[44] Arminjon M, Reifler F. Equivalent forms of Dirac equations in curved spacetimes and generalized de Broglie relations. Brazilian Journal of Physics. 2013;**43**(1–2):64-77

[45] Arminjon M, Reifler F. Basic quantum mechanics for three dirac equations in a curved spacetime. Brazilian Journal of Physics. 2010;**40**(2): 242-255 arXiv:0807.0570

[46] Pollock MD. On the Dirac equation in curved spacetime (Moscow, Russia). Acta Physica Polonica B. 2010;**41**(8): 1827-1846

[47] Arminjon M. Dirac-type equations in a gravitational field, with vector wave function. Foundations of Physics. 2008; **38**(11):1020-1045

[48] Parashar D. Dirac equation in curved space. Il Nuovo Cimento A (1965-1970). 1990;**103**(10):1491-1493. DOI: 10.1007/BF02820576

[49] Weyl HKH. Elektron und gravitation I. Zeitschrift für Physik. 1927;**56**:330-352

[50] Weyl HKH. Gravitation and the Electron. Proceedings of the National Academy of Sciences of the United States of America. 1927;**15**:323-334

[51] Fock VA. Geometrization of the Dirac theory of electrons. Zeitschrift für Physik. 1929;**57**:261-277

[52] Robson BA. The Matter-antimatter asymmetry problem. 2018. DOI: 10.4236/jhepgc.2018.41015

[53] Robson B. The generation model of particle physics. In: Particle Physics. InTech; 2012. DOI: 10.5772/35071

[54] Robson BA. The generation model and the origin of mass. International Journal of Modern Physics E. 2009; **18**(08):1773-1780. DOI: 10.1142/s0218301309013786

[55] Robson BA. A quantum theory of gravity based on a composite model of

leptons and quarks. International Journal of Modern Physics E. 2011; **20**(03):733-745. DOI: 10.1142/s0218301311018198

[56] Evans PW, Robson BA. Comparison of quark mixing in the standard and generation models. International Journal of Modern Physics E. 2006;**15**(03):17-625. DOI: 10.1142/s0218301306004077

[57] Robson BA. A generation model of composite leptons and quarks. International Journal of Modern Physics E. 2005;**14**(08):1151-1169. DOI: 10.1142/s0218301305003776

[58] Robson BA. Relation between Strong and weak isospin. International Journal of Modern Physics E. 2004; **13**(05):999-1018. DOI: 10.1142/s0218301304002521

[59] Robson BA. A generation model of the fundamental particles. International Journal of Modern Physics E. 2002; **11**(06):555-566. DOI: 10.1142/s0218301302001125

[60] Heras JA. Can Maxwell's equations Be obtained from the continuity equation? American Journal of Physics. 2007;**75**:652. DOI: 10.1119/1.2739570

[61] Behera H. Newtonian Gravitomagnetism and Analysis of Earth Satellite Results. arXiv:Grqc/0510003v2; 2006

[62] Bohm D. A suggested interpretation of the quantum theory in terms of 'hidden variables' I. Physics Review. 1952;**84**:166-179. DOI: 10.1103/physrev.85.166

[63] Bohm D. A suggested interpretation of the quantum theory in terms of 'hidden variables' II. Physics Review. 1952;**85**:180-193. DOI: 10.1103/physrev.85.180

[64] Bohm D. Proof that probability density approaches $|\psi|^2$ in causal interpretation of the quantum theory. Physical Review. 1953;**89**(2):458-466. DOI: 10.1103/physrev.89.458